ISBN 978-1-330-57701-1
PIBN 10006491

1 MONTH OF
FREE
READING

at

www.ForgottenBooks.com

By purchasing this book you are eligible for one month membership to ForgottenBooks.com, giving you unlimited access to our entire collection of over 700,000 titles via our web site and mobile apps.

To claim your free month visit:

www.forgottenbooks.com/free6491

Similar Books Are Available from
www.forgottenbooks.com

What I Know of Farming:

A SERIES OF

BRIEF AND PLAIN EXPOSITIONS

OF

PRACTICAL AGRICULTURE

AS AN ART BASED UPON SCIENCE:

By HORACE GREELEY.

"*I know*
That where the spade is deepest driven,
The best fruits grow."
JOHN G. WHITTIER.

NEW-YORK:
PUBLISHED BY THE TRIBUNE ASSOCIATION.
1871.

TO

THE MAN OF OUR AGE,

WHO SHALL MAKE THE FIRST PLOW PROPELLED BY

STEAM,

OR OTHER MECHANICAL POWER, WHEREBY NOT LESS THAN

TEN ACRES PER DAY

SHALL BE THOROUGHLY PULVERIZED TO A

DEPTH OF TWO FEET,

AT A COST OF NOT MORE THAN TWO DOLLARS PER ACRE,

THIS WORK IS ADMIRINGLY DEDICATED BY

THE AUTHOR.

CONTENTS BY CHAPTERS.

PREFACE.

————

MEN have written wisely and usefully, in illustration and aid of Agriculture, from the platform of pure science. Acquainted with the laws of vegetable growth and life, they so expounded and elucidated those laws that farmers apprehended and profitably obeyed them. Others have written, to equally good purpose, who knew little of science, but were adepts in practical agriculture, according to the maxims and usages of those who have successfully followed and dignified the farmer's calling. I rank with neither of these honored classes. My practical knowledge of agriculture is meager, and mainly acquired in a childhood long bygone; while, of science, I have but a smattering, if even that. They are right, therefore, who urge that my qualifications for writing on agriculture are slender indeed.

I only lay claim to an invincible willingness to be made wiser to-day than I was yesterday, and a lively faith in the possibility—nay, the feasibility, the urgent necessity, the imminence—of very great improvements in our ordinary dealings with the soil. I know that a majority of those who would live by its tillage feed it too sparingly and stir it too

7

slightly and grudgingly. I know that we do too little for it, and expect it, thereupon, to do too much for us. I know that, in other pursuits, it is only work thoroughly well done that is liberally compensated; and I see no reason why farming should prove an exception to this stern but salutary law. I may be, indeed, deficient in knowledge of what constitutes good farming, but not in faith that the very best farming is that which is morally sure of the largest and most certain reward.

I hope to be generally accorded the merit of having set forth the little I pretend to know in language that few can fail to understand. I have avoided, so far as I could, the use of terms and distinctions unfamiliar to the general ear. The little I know of oxygen, hydrogen, nitrogen, &c., I have kept to myself; since whatever I might say of them would be useless to those already acquainted with the elementary truths of Chemistry, and only perplexing to others. If there is a paragraph in the following pages which will not be readily and fully understood by an average school-boy of fifteen years, then I have failed to make that paragraph as simple and lucid as I intended.

Many farmers are dissuaded from following the suggestions of writers on agriculture by the consideration of expense. They urge that, though men of large wealth may (perhaps) profitably do what is recommended, *their* means are utterly inadequate: they might as well be urged to work their oxen in a silver yoke with gold bows. I have aimed to commend mainly, if not uniformly, such improvements only on our grandfathers' husbandry as a farmer worth $1,000, or over, may adopt—not all at once, but gradually, and from year to year. I hope I shall thus convince some farmers that draining, irrigation, deep plowing, heavy fertilizing, &c., are not beyond their power, as so many have too readily presumed and pronounced them.

That I should say very little, and that little vaguely, of the breeding and raising of animals, the proper time to sow or plant, &c., &c., can need no explanation. By far the larger number of those whose days have mainly been given to farming, know more than I do of these details, and are better authority than I am with regard to them. On the other hand, I have traveled extensively, and not heedlessly, and have seen and pondered certain broader features of the earth's improvement and tillage which many stay-at-home cultivators have had little or no opportunity to study or even observe. By restricting the topics with which I deal, the probability of treating some of them to the average farmer's profit is increased.

And, whatever may be his judgment on this slight work, I *know* that, if I could have perused one of like tenor half a century ago, when I was a patient worker and an eager reader in my father's humble home, my subsequent career would have been less anxious and my labors less exhausting than they have been. Could I then have caught but a glimpse of the beneficent possibilities of a farmer's life—could I have realized that he is habitually (even though blindly) dealing with problems which require and reward the amplest knowledge of Nature's laws, the fullest command of science, the noblest efforts of the human intellect, I should have since pursued the peaceful, unobtrusive round of an enthusiastic and devoted, even though not an eminent or fortunate, tiller of the soil. Even the little that is unfolded in the ensuing pages would have sufficed to give me a far larger, truer, nobler conception of what the farmer of moderate means might and should be, than I then attained. I needed to realize that observation and reflection, study and mental acquisition, are as essential and as serviceable in his pursuit as in others, and that no man can have acquired so much general knowledge that a farmer's exigencies will not afford scope and use for it all. I abandoned the farm, because I fancied that I had already perceived, if I had not as yet clearly comprehended, all there was in the farmer's calling; whereas, I had not really learned much more of it than a good plow-horse ought to understand. And, though great progress has been made since then, there are still thousands of boys, in this enlightened age and conceited generation, who have scarcely a more adequate and just conception of agriculture than I then had. If I could hope to reach even one in every hundred of this class, and induce him to ponder, impartially, the contents of this slight volume, I know that I shall not have written it in vain.

We need to mingle more thought with our work. Some think till their heads ache intensely; others work till their backs are crooked to the semblance of half an iron hoop; but the workers and the thinkers are apt to be distinct classes; whereas, they should be the same. Admit that it has always been thus, it by no means follows that it always should or shall be. In an age when every laborer's son may be fairly educated_ if he will, there should be more fruit gathered from the tree of knowledge to justify the magnificent promise of its foliage and its bloom. I rejoice in the belief that the graduates of our common schools are better ditch-diggers, when they can no otherwise employ themselves to better advantage, than though they knew not how to read; but that is not enough. If the untaught peasantry of Russia or Hungary grow more wheat per acre than the comparatively educated farmers of the United States, our education is found wanting. That is a vicious and defective if not radically false mental training which leaves its subject no better qualified for any useful calling than though he were unlettered. But I forbear to pursue this ever-fruitful theme.

I look back, on this day completing my sixtieth year, over a life, which must now be near its close, of constant effort to achieve ends whereof many seem in the long retrospect to have been transitory and unimportant, however they may have loomed upon my vision when in their immediate presence. One achievement only of our age and country—the banishment of human chattelhood from our soil—seems now to have been worth all the requisite efforts, the agony and bloody sweat,.through which it was accomplished. But another reform, not so palpably demanded by justice and humanity, yet equally conducive to the well-being of our race, presses hard on its heels, and insists that we shall accord it instant and earnest consideration. It is the elevation of Labor from the plane of drudgery and servility to one of self-respect, self-guidance, and genuine independence, so as to render the human worker no mere cog in a vast, revolving wheel, whose motion he can neither modify nor arrest, but a partner in the enterprise which his toil is freely contributed to promote, a sharer in the outlay, the risk, the loss and gain, which it involves. This end can be attained through the training of the generation who are to succeed us to observe and reflect, to live for other and higher ends than those of present sensual gratification, and to feel that no achievement is beyond the reach of their wisely combined and ably self-directed efforts. To that part of the generation of farmers just coming upon the stage of responsible action, who have intelligently resolved that the future of American agriculture shall evince decided and continuous improvement on its past, this little book is respectfully commended.   H. G.

New York, Feb. 8, 1871.

WHAT I KNOW OF FARMING.

I.

I COMMENCE my essays with this question, because, when I urge the superior advantages of a rural life, I am often met by the objection that *Farming does n't pay.* That, if true, is a serious matter. Let us consider:

I do not understand it to be urged that the farmer who owns a large, fertile estate, well-fenced, well-stocked, with good store of effective implements, cannot live and thrive by farming. What is meant is, that he who has little but two brown hands to depend upon cannot make money, or can make very little, by farming.

I think those who urge this point have a very inadequate conception of the difficulty encountered by every poor young man in securing a good start in life, no matter in what pursuit. I came to New-York when not quite of age, with a good constitution, a fair common-school education, good health, good habits, and a pretty fair trade—(that of printer.) I think

(13)

my outfit for a campaign against adverse fortune was decidedly better than the average; yet ten long years elapsed before it was settled that I could remain here and make any decided headway. Meantime, I drank no liquors, used no tobacco, attended no balls or other expensive entertainments, worked hard and long whenever I could find work to do, lost less than a month altogether by sickness, and did very little in the way of helping others. I judge that quite as many did worse than I as did better; and that, of the young lawyers and doctors who try to establish themselves here in their professions, quite as many earn less as earn more than their bare board during the first ten years of their struggle.

John Jacob Astor, near the close of a long, diligent, prosperous career, wherein he amassed a large fortune, is said to have remarked that, if he were to begin life again, and had to choose between making his first thousand dollars with nothing to start on, or with that thousand making all that he had actually accumulated, he would deem the latter the easier task. Depend upon it, young men, it is and must be hard work to earn honestly your first thousand dollars. The burglar, the forger, the blackleg (whether he play with cards, with dice, or with stocks), may seem to have a quick and easy way of making a thousand dollars; but whoever makes that sum honestly, with nothing but his own capacities and energies as capital, does a very good five-years' work, and may deem himself fortunate if he finishes it so soon.

I *have* known men do better, even at farming. I recollect one who, with no capital but a good wife and four or five hundred dollars, bought (near Boston) a farm of two hundred mainly rough acres, for $2,500, and paid for it out of its products within the next five years, during which he had nearly doubled its value. I lost sight of him then; but I have not a doubt that, if he lived fifteen years longer and had no very bad luck, he was worth, as the net result of twenty years' effort, at least $100,000. But this man would rise at four o'clock of a winter morning, harness his span of horses and hitch them to his large market-wagon (loaded over night), drive ten miles into Boston, unload and load back again, be home at fair breakfast-time, and, hastily swallowing his meal, be fresh as a daisy for his day's work, in which he would lead his hired men, keeping them clear of the least danger of falling asleep. Such men are rare, but they still exist, proving scarcely anything impossible to an indomitable will. I would not advise any to work so unmercifully; I seek only to enforce the truth that great achievements are within the reach of whoever will pay their price.

An energetic farmer bought, some twenty-five years ago, a large grazing farm in Northern Vermont, consisting of some 150 acres, and costing him about $3,000. He had a small stock of cattle, which was all his land would carry; but he resolved to increase that stock by at least ten per cent. per annum, and to so improve his land by cultivation, fertilizing, clover, &c., that it would amply carry that increase. Fifteen years later, he sold out farm and stock for $45,000, and migrated to the West. I did not understand that he was a specially hard worker, but only a good manager, who kept his eyes wide open, let nothing go to waste, and steadily devoted his energies and means to the improvement of his stock and his farm.

Walking one day over the farm of the late Prof. Mapes, he showed me a field of rather less than ten acres, and said, " I bought that field for $2,400, a year ago last September. There was then a light crop of corn on it, which the seller reserved and took away. I underdrained the field that Fall, plowed and sub-soiled it, fertilized it liberally, and planted it with cabbage; and, when these matured, I sold them for enough to pay for land, labor, and fertilizers, altogether." The field was now worth far more than when he bought it, and he had cleared it within fifteen months from the date of its purchase. I consider that a good operation. Another year, the crop might have been poor, or might have sold much lower, so as hardly to pay for the labor; but there are risks in other pursuits as well as in farming.

A fruit-farmer, on the Hudson above Newburg, showed me, three years since, a field of eight or ten acres which he had nicely set with Grapes, in rows ten feet apart, with beds of Strawberries between the rows, from which he assured me that his sales per acre exceeded $700 per annum. I presume his outlay for labor, including picking, was less than $300 per annum; but it had cost something to make this field what it then was. Say that he had spent $1,000 per acre in under-draining, enriching and tilling this field, to bring it to this condition, including the cost of his plants, and still there must have been a clear profit here of at least $300 per acre.

I might multiply illustrations; but let the foregoing suffice. I readily admit that shiftless farming does n't pay—that poor crops do n't pay—that it is hard work to make money by farming without some capital—that frost, or hail, or drouth, or floods, or insects, may blast the farmer's hopes, after he has done his best to deserve and achieve success; but I insist that, as a general proposition, GOOD *Farming* DOES *pay*—that few pursuits afford as good a prospect, as full an assurance, of reward for intelligent, energetic, persistent effort, as this does.

I am not arguing that every man should be a farmer. Other vocations are useful and necessary, and many pursue them with advantage to themselves and to others. But those pursuits are apt to be modified by time, and some of them may yet be entirely dispensed with, which Farming never can be. It is the first and most essential of human pursuits; it is every one's interest that this calling should be honored and prosperous. If not adequately recompensed, I judge that is because it is not wisely and energetically followed. My aim is to show how it may be pursued with satisfaction and profit.

II

GOOD AND BAD HUSBANDRY.

NECESSITY is the master of us all. A farmer may be as strenuous for deep plowing as I am—may firmly believe that the soil should be thoroughly broken up and pulverized to a depth of fifteen to thirty inches, according to the crop; but, if all the team he can muster is a yoke of thin, light steers, or a span of old, spavined horses, which have not even a speaking acquaintance with grain, what shall he do? So he may heartily wish he had a thousand loads of barn-yard manure, and know how to make a good use of every ounce of it; but, if he has it not, and is not able to buy it, he can't always afford to forbear sowing and planting, and so, because he cannot secure great crops, do without any crops at all. If he does the best he can, what better *can* he do?

Again: Many farmers have fields that must await the pleasure of Nature to fit them for thorough cultivation. Here is a field—sometimes a whole farm—which, if partially divested of the primitive forest, is still thickly dotted with obstinate stumps and filled with green, tenacious roots, which could only be re-

(18)

moved at a heavy, perhaps ruinous, cost. A rich man might order them all dug out in a month and see his order fully obeyed; but, except to clear a spot for a garden or under very peculiar circumstances, it would not pay; and a poor man cannot afford to incur a heavy expense merely for appearance's sake, or to make a theatrical display of energy. In the great majority of cases, he who farms for a living can't afford to pull green stumps, but must put his newly-cleared land into grass at the earliest day, mow the smoother, pasture the rougher portions of it, and wait for rain and drouth, heat and frost, to rot his stumps until they can easily be pulled or burned out as they stand.

So with regard to a process I detest, known as Pasturing. I do firmly believe that the time is at hand when nearly all the food of cattle will, in our Eastern and Middle States, be cut and fed to them—that we can't afford much longer, even if we can at present, to let them roam at will over hill and dale, through meadow and forest, biting off the better plants and letting the worse go to seed; often poaching up the soft, wet soil, especially in Spring, so that their hoofs destroy as much as they eat; nipping and often killing in their infancy the finest trees, such as the Sugar Maple, and leaving only such as Hemlock, Red Oak, Beech, &c., to attain maturity. Our race generally emerged from savageism and squalor into industry, comfort and thrift, through the Pastoral condition—the herding, taming, rearing and training

of animals being that department of husbandry to
which barbarians are most easily attracted: hence,
we cling to Pasturing long after the reason for it has
vanished. The radical, incurable vice of Pasturing—
that of devouring the better plants and leaving the
worse to ripen and diffuse seed—can never be wholly
obviated; and I deem it safe to estimate that almost
any farm will carry twice as much stock if their food
be mainly cut and fed to them as it will if they are
required to pick it up where and as it grows or grew.
I am sure that the general adoption of Soiling instead
of Pasturing will add immensely to the annual product, to the wealth, and to the population, of our
older States.   And yet, I know right well that many
farms are now so rough and otherwise so unsuited to
Soiling as to preclude its adoption thereon for many
years to come.

Let me indicate what I mean by Good Farming,
through an illustration drawn from the Great West:

All over the settled portions of the Valley of the
Upper Mississippi and the Missouri, there are large
and small herds of cattle that are provided with little
or no shelter. The lea of a fence or stack, the partial protection of a young and leafless wood, they
may chance to enjoy; but that it is a ruinous waste
to leave them a prey to biting frosts and piercing
north-westers, their owners seem not to comprehend.
Many farmers far above want will this Winter feed
out fields of Corn and stacks of Hay to herds of cattle that will not be one pound heavier on the 1st of

next May than they were on the 1st of last December —who will have required that fodder merely to preserve their vitality and escape freezing to death. It has mainly been employed as fuel rather than as nourishment, and has served, not to put on flesh, but to keep out frost.

Now I am familiar with the excuses for this waste; but they do not satisfy me. The poorest pioneer might have built for his one cow a rude shelter of stakes, and poles, and straw or prairie-grass, if he had realized its importance, simply in the light of economy. He who has many cattle is rarely without both straw and timber, and might shelter his stock abundantly if he only would. Nay, he could not have neglected or omitted it if he had clearly understood that his beasts must somehow be supplied with heat, and that he can far cheaper warm them from without than from within.

The broad, general, unquestionable truths, on which I insist in behalf of Good Farming are these; and I do not admit that they are subject to exception :

I. It is very rarely impracticable to grow good crops, if you are willing to work for them. If your land is too poor to grow Wheat or Corn, and you are not yet able to enrich it, sow Rye or Buckwheat; if you cannot coax it to grow a good crop of anything, let it alone; and, if you cannot run away from it, work out by the day or month for your more fortunate neighbors. The time and means squandered in trying to grow crops where only half or quarter crops can be made, constitute the heaviest item on the wrong side of our farmers' balance-sheets; taxing them more than their National, State, and local governments together do.

II. Good crops rarely fail to yield a profit to the grower. I know there are exceptions, but they are very few. Keep your eye on the farmer who almost uniformly has great Grass, good Wheat, heavy Corn, &c., and, unless he drinks, or has some other bad habit, you will find him growing rich. I am confident that white blackbirds are nearly as abundant as farmers who have become poor while usually growing good crops.

III. The fairest single test of good farming is the increasing productiveness of the soil. That farm which averaged twenty bushels of grain to the acre twenty years ago, twenty-five bushels ten years ago, and will measure up thirty bushels to the acre from this year's crop, has been and is in good hands. I know no other touchstone of Farming so unerring as that of the increase or decrease from year to year of its aggregate product. If you would convince me that X. is a good farmer, do not tell me of some great crop he has just grown, but show me that his crop has regularly increased from year to year, and I am satisfied.

—I shall have more to say on these points as I proceed. It suffices for the present if I have clearly indicated what I mean by Good and what by Bad Farming.

III.

WHEN my father was over sixty years old, and had lived some twenty years in Erie County, Pennsylvania, he said to me· "I have several times removed, and always toward the West; I shall never remove again; but, were I to do so, it would be toward the East. Experience has taught me that the advantages of every section are counterbalanced by disadvantages, and that, where any crop is easily produced, there it sells low, and sometimes cannot be sold at all. I shall live and die right here; but, were I to remove again, it would not be toward the West."

This is but one side of a truth, and I give it for whatever it may be worth. Had my father plunged into the primitive forest in his twenty-fifth rather than his forty-fifth year, he would doubtless have become more reconciled to pioneer life than he ever did. I would advise no one over forty years of age to undertake, with scanty means, to dig a farm out of the dense forest, where great trees must be cut down and cut up, rolled into log-heaps, and burned to ashes where they grew. Where half the timber can be

(23)

sold for enough to pay the cost of cutting, the case is different ; but I know right well that digging a farm out of the high woods is, to any but a man of wealth, a slow, hard task. Making one out of naked prairie, five to ten miles from timber, is less difficult, but not much. He who can locate where he has good timber on one side and rich prairie on the other is fortunate, and may hope, if his health be spared, to surround himself with every needed comfort within ten years. Still, the pioneer's life is a rugged one, especially for women and children ; and I should advise any man who is worth $2,000 and has a family, to buy out an "improvement" (which, in most cases, badly needs improving) on the outskirt of civilization, rather than plunge into the pathless forest or push out upon the unbroken prairie. I rejoice that our Public Lands are free to actual settlers ; I believe that many are thereby enabled to make for themselves homes who otherwise would have nothing to leave their children ; yet I much prefer a home within the boundaries of civilization to one clearly beyond them. There is a class of drinking, hunting, frolicking, rarely working, frontiersmen, who seem to have been created on purpose to erect log cabins and break paths in advance of a different class of settlers, who regularly come in to buy them out and start them along after a few years. I should here prefer to follow rather than lead. If Co-operation shall ever be successfully applied to the improvement of wild lands, I trust it may be otherwise.

He who has a farm already, and is content with it, has no reason to ask, " Whither shall I go ?" and he may rest assured that thoroughly good farming will pay as well in New England as in Kansas or in Minnesota. I advise no man who has a good farm anywhere, and is able to keep it, to sell and migrate. I know men who make money by growing food within twenty miles of this city quite as fast as they could in the West. If you have money to buy and work it, and know how to make the most of it, I believe you may find land really as cheap, all things considered, in Vermont as in Wisconsin or Arkansas.

And yet I believe in migration—believe that there are thousands in the Eastern and the Middle States who would improve their circumstances and prospects by migrating to the cheaper lands and broader opportunities of the West and South. For, in the first place, most men are by migration rendered more energetic and aspiring ; thrown among strangers, they feel the necessity of exertion as they never felt it before. Needing almost everything, and obliged to rely wholly ou themselves, they work in their new homes as they never did in their old ; and the cousequences are soon visible all around them.

" A stern chase is a long chase," say the sailors ; and he who buys a farm mainly on credit, intending to pay for it out of its proceeds, finds interest, taxes, sickness, bad seasons, hail, frost, drouth, tornadoes, floods, &c., &c., deranging his calculations and impeding his progress, until he is often impelled to

2

give up in despair. There are men who can surmount every obstacle and defy discouragement—
these need no advice; but there are thousands who,
having little means and large families, can grow into
a good farm more easily and far more surely than
they can pay for it; and these may wisely seek homes
where population is yet sparse and land is consequently cheap. Doubtless, some migrate who might better have forborne; yet the instinct which draws our
race toward sunset is nevertheless a true one. The
East will not be depopulated; but the West will grow
more rapidly in the course of the next twenty years
than ever in the past. The Railroads which have
brought Kansas and Minnesota within three days,
and California within a week of us, have rendered
this inevitable.

But the South also invites immigration as she never
did till now. Her lands are still very cheap; she is
better timbered, in the average, than the West; her
climate attracts; her unopened mines and unused
water-power call loudly for enterprise, labor and
skill. It is absurd to insist that her soil is exhausted
when not one-third of it has ever yet been plowed.
I do not advise solitary migration to the South, because she needs schools, mills, roads, bridges,
churches, &c., &c., which the solitary immigrant can
neither provide nor well do without: and I have no
assurance that he, if obliged to work out for present
bread, would find those ready to employ and willing
to pay him; but let a hundred Northern farmers and

mechanics worth $1,000 to $3,000 each combine to select (through chosen agents) and buy ten or twenty thousand acres in some Southern State, embracing hill and vale, timber and tillage, water-power and minerals, and divide it equitably among themselves, after laying it out with roads, a park, a village-plat, sites for churches, schools, &c., and I am confident that they can thus make pleasant homes more cheaply and speedily there than almost anywhere else.

Good farming land, improved or unimproved, is this day cheaper in the United States, all things considered, than in any other country—cheaper than it can long remain. So many are intent on short cuts to riches that the soil is generally neglected, and may be bought amazingly cheap in parts of Connecticut as well as in Iowa or Nebraska. When I was last in Illinois, I rode for some hours beside a gray-coated farmer of some sixty years, who told me this· "I came here thirty years ago, and took up, at $1¼ per acre, a good tract of land, mainly in timber. I am now selling off the timber at $100 per acre, reserving the land." That seems to me a good operation— not so quick as a corner in the stock-market, but far safer. And, while I would advise no man to incur debt, I say most earnestly to all who have means, "Look out the place where you would prefer to live and die; take time to suit yourself thoroughly; choose it with reference to your means, your calling, your expectations, and, if you can pay for it, *buy* it. Do not imagine that land is cheap in the West or

the South only ; it is to be found cheap in *every*
State by those who are able to own and who know
how to use it."

I earnestly trust that the obvious advantages of
settling in colonies are to be widely and rapidly improved by our people, nearly as follows :   One thous-
and heads of families unite to form a colony, contribute
$100 to $500 each to defray the cost of seeking out
and securing a suitable location, and send out two or
three of the most capable and trustworthy of their number to find and purchase it ; and now let their lands be
surveyed and divided into village or city lots at or near
the center, larger allotments (for mechanics' and merchants' homes) surrounding that center, and far larger
(for farms) outside of these ; and let each member, on
or soon after his arrival, select a village-lot, out-lot,
farm, or one of each if he chooses and can pay for
them.   Let ample reservations of the best sites for
churches, school-houses, a town hall, public park, etc.,
be made in laying out the village, and let each pur-
chaser of a lot or farm be required to plant shade-trees
along the highways which skirt or traverse it.   If
irrigation by common effort be deemed necessary, let
provision be made for that.   Run up a large, roomy
structure for a family hotel or boarding-house ; and
now invite each stockholder to come on, select his land,
pay for it, and get up some sort of a dwelling, leaving
his family to follow when this shall have been rendered
habitable ; but, if they insist on coming on with him
and taking their chances, so be it.

IV

I write mainly for beginners—for young persons, and some not so young, who are looking to farming as the vocation to which their future years are to be given, by which their living is to be gained. In this chapter, I would counsel young men, who, not having been reared in personal contact with the daily and yearly round of a farmer's cares and duties, purpose henceforth to live by farming.

To these I would earnestly say, "No haste!" Our boys are in too great a hurry to be men. They want to be bosses before they have qualified themselves to be efficient journeymen. I have personally known several instances of young men, fresh from school or from some city vocation, buying or hiring a farm and undertaking to work it; and I cannot now recall a single instance in which the attempt has succeeded; while speedy failure has been the usual result. The assumption that farming is a rude, simple matter, requiring little intellect and less experience, has buried many a well-meaning youth under debts which the best efforts of many subsequent years will barely enable him to pay off. In my opinion, half our farmers now living would say, if questioned, that they might better have waited longer before buying or hiring a farm.

When I was ten years old, my father took a job of clearing off the mainly fallen and partially rotten timber—largely White Pine and Black Ash—from fifty acres of level and then swampy land; and he and his two boys gave most of the two ensuing years (1821–2) to the rugged task. When it was finished, I—a boy of twelve years—could have taken just such a tract of half-burned primitive forest as that was when we took hold of it, and cleared it by an expenditure of seventy to eighty per cent. of the labor we actually bestowed upon that. I had learned, in clearing this, how to economize labor in any future undertaking of the kind; and so every one learns by experience who steadily observes and reflects. He must have been a very good farmer at the start, or a very poor one afterward, who cannot grow a thousand bushels of grain much cheaper at thirty years of age than he could at twenty.

To every young man who has had no farming experience, or very little, yet who means to make farming his vocation, I say, Hire out, for the coming year, to the very best farmer who will give you anything like the value of your labor. Buy a very few choice books, (if you have them not already,) which treat of Geology, Chemistry, Botany, and the application of their truths in Practical Agriculture; give to these the close and thoughtful attention of your few leisure hours; keep your eyes wide open, and set down in a note-book or pocket-diary each night a minute of whatever has been done on the farm that day, making a note of each storm, shower, frost, hail, etc., and also of the date at which each planted crop requires tillage or is ripe enough to harvest, and ascertaining, so far as possible, what each crop produced on the farm has cost, and which of them all are produced at a profit and which at a loss. At the year's end, hire again to the same or another good farmer and pursue the same course; and so do till you shall be twenty-four or twenty-five years of age, which is young enough to marry, and quite young enough to under-take the management· of a farm. By this time, if you have carefully saved and wisely invested your earnings, you will have several hundred dollars; and, if you do not choose to migrate to some region where land is very cheap, you will have found some one willing to sell you a small farm on credit, taking a long mortgage as security. Your money—assuming that you have only what you will have earned—will all be wanted to fix up your building, buy a team and cow, with the few implements needed, and supply you with provisions till you can grow some. If you can start thus experienced and full-handed, you may, by diligence, combined with good fortune, begin to make payments on your mortgage at the close of your second year.

I hate debt as profoundly as any one can, but I do

not consider this really running into debt. One has
more land than he needs, and does not need his pay
for it forthwith ; another wants land, but lacks the
means of present payment. They two enter into an
agreement mutually advantageous, whereby the poorer has the present use and ultimate fee-simple of the
farm in question, in consideration of the payment of
certain sums as duly stipulated. Technically, the
buyer becomes a debtor ; practically, I do not regard
him as such, until payments fall due which he is unable promptly to meet. Let him rigorously avoid
all other debt, and he need not shrink from nor be
ashamed of this.

I have a high regard for scientific attainments ; I
wish every young man were thoroughly instructed in
the sciences which underlie the art of farming. But
all the learning on earth, though it may powerfully
help to make a good farmer, would not of itself make
one. When a young man has learned all that semi-
naries and lectures, books and cabinets, can teach
him, he still needs practice and experience to make
him a good farmer.

—"But wouldn't you have a young man study in
order that he may become a good farmer ?"

—If he has money, Yes. I believe a youth worth
four or five thousand dollars may wisely spend a
tenth of his means in attending lectures, and even
courses of study, at any good seminary where Natural
Science is taught and applied to Agriculture. But
life is short at best ; and he who has no means, or

very little, cannot really afford to attend even an Agricultural College. He can acquire so much of Science as is indispensable in the cheaper way I have indicated. He cannot wisely consent to spend the best years of his life in getting ready to live.

He who has already mastered the art of farming, and has adequate means, may of course buy a farm to-morrow, though he be barely or not quite of age. He has little to learn from me. Yet I think even such have often concluded, in after years, that they were too hasty in buying land—that they might profitably have waited, and deliberated, and garnered the treasures of experience, before they took the grave step of buying their future home; with regard to which I shall make some suggestions in my next chapter.

But I protest against a young man's declining or postponing the purchase of a farm merely because he is not able to buy a great one. Twenty acres of arable soil near a city or manufacturing village, forty acres in a rural district of any old State, or eighty acres in a region just beginning to be peopled by White men, is an ample area for any one who is worth less than $2,000. If he understands his business, he will find profitable employment hereon for every working hour: if he does not understand farming, he will buy his experience dear enough on this, yet more cheaply than he would on a wider area. Until he shall have more money than he needs, let him beware of buying more land than he absolutely wants.

2*

V.

No one need be told at this day that good land is cheaper than poor—that the former may be bought at less cost than it can be made. Yet this, like most truths, may be given undue emphasis. It should be considered in the light of the less obvious truth that *Every farmer may make advantageous use of* SOME *poor land.* The smallest farm should have its strip or belt of forest; the larger should have an abundance and variety of trees; and sterile, stony land grows many if not most trees thriftily. Even at the risk of arousing Western prejudice, I maintain that New-England, and all broken, hilly, rocky countries, have a decided advantage (abundantly counterbalanced, no doubt) over regions of great fertility and nearly uniform facility, in that human stupidity and mole-eyed greed can never wholly divest them of forests—that their sterile crags and steep acclivities must mainly be left to wood forever. Avarice may strip them of their covering of to-day; but, defying the plow and the spade, they cannot be so denuded that they will not be speedily rëclothed with trees and foliage.

(34)

I am not a believer that "Five Acres" or "Ten Acres" suffice for a farm. I know where money is made on even fewer than five acres; but they who do it are few, and men of exceptional capacity and diligence. Their achievements are necessarily confined to the vicinage of cities or manufacturing villages. The great majority of all who live by Agriculture want room to turn upon—want to grow grass and keep stock—and, for such, no mere garden or potato-patch will answer. They want genuine farms.

Yet, go where you may in this country, you will hear a farmer saying of his neighbor, "He has too much land," even where the criticism might justly be reciprocated. We cannot all be mistaken on this head.

There are men who can each manage thousands of acres of tillage, just as there are those who can skillfully wield an army of a hundred thousand men. Napoleon said there were two of this class in the Europe of his day. There are others who cannot handle a hundred acres so that nothing is lost through neglect or oversight. Rules must be adapted to average capacities and circumstances. He who expects to live by cattle-rearing needs many more acres than he who is intent on grain-growing; while he who contemplates vegetable, root, and fruit culture, needs fewer acres still. As to the direction of his efforts, each one will be a law unto himself.

If I were asked, by a young man intent on farming, to indicate the proper area for him, I would say, *Buy just so large a farm as half your means will pay for.* In other words, " If you are worth $20,000, invest half of it in land, the residue in stock, tools, etc. ; and observe the same rule of proportion, whether you be worth $1,000,000 or only $1,000. If you are worth just nothing at all, I would invest in land the half of that, and no more. In other words, I would either wait to earn $500 or over, or push Westward till I found land that costs practically nothing.

This, then, I take to be the gist of the popular criticism on our farmers as having unduly enlarged their borders : *They have more land than they have capital to stock and till to the best advantage.* He who has but fifty acres has too much if he lets part of his land lie idle and unproductive for lack of team or hands to till it efficiently ; while he who has a thousand acres has none too much if he has the means and talents wherewith to make the best of it all.

I have said that I consider the soil of New England .as cheap, all things considered, for him who is able to buy and work it, as that of Minnesota or Arkansas—that I urge migration to the West only upon those who cannot pay for farms in the old States. I doubt whether the farmers of any other section have, in the average, done better, throughout the last ten years, than the butter-makers of Vermont, the cheese-dairymen of this State. And yet there is, in the ridgy, rocky, *patchy* character of most of our Eastern farms, an insuperable barrier to the most economic, effective cultivation. If the ridges were further apart—if each rocky or gravelly knoll were not in close proximity to a strip of bog or morass—it would be different. But the genius of our age points unmistakably to cultivation by steam or some other mechanical application of power; and this requires spacious fields, with few or no obstacles to the equable progress of the plow. I apprehend that, for this reason, the growth of bread-corn eastward of the Hudson can never more be considerably extended, so long as the boundless, fertile prairies can so easily pour their exhaustless supplies upon us. Fruits, Vegetables, Roots and Grass, we must continue to grow, probably in ever-increasing abundance; but we of the East will buy our bread-corn largely if not mainly from the West.

He, therefore, who buys land in the Eastern States should regard primarily its capacity to produce those crops in which the East can never be supplanted— Grass, Fruits, Vegetables, Timber. If a farm will also produce good Corn or Wheat, that is a recommendation; but let him place a higher value on those capacities which will be more generally required and drawn upon.

In the West, the case is different; for, though Wheat-culture still recedes before the footsteps of advancing population, and Minnesota may soon cease to grow for others, as Western New-York, Ohio, Indiana, and Northern Illinois, have already done, yet Indian Corn, being the basis of both Beef and Pork, will long hold its own in the Valley of the Ohio and in that of the Upper Mississippi. As it recedes slowly Westward, Clover and Timothy, Butter and Cheese, will press closely on its footsteps.

Good neighbors, good roads, good schools, good mechanics at hand, and a good church within reach, will always be valued and sought : few farmers are likely to disregard them. Let whoever buys a farm whereon to live resolve to buy once for all, and let him not forget that health is not only wealth but happiness—that an eligible location and a beautiful prospect are elements of enjoyment not only for ourselves but our friends ; let him not fancy that all the land will soon be gobbled up and held at exorbitant prices, but believe that money will almost always command money's worth of whatever may be needed, so that he need not embarrass himself to-day through fear that he may not be able to find sellers to-morrow, and he can hardly fail to buy judiciously, and thus escape that worst species of home-sickness— sickness of home.

VI.

WHOEVER finds himself the newly installed owner and occupant of a farm, should, before doing much beyond growing a crop in the ordinary way, study well its character, determine its capacities, make himself well acquainted with its peculiarities of soil and surface, with intent to make the most of it in his future operations. I would devote at least a year to this thoughtful observation and study.

To one reared amid the rugged scenery of New-England, or on either slope of the Allegheny ridge, all prairie farms look alike, just as a European supposes this to be the case with all negroes. A better acquaintance will show the average prairie quarter-section by no means an unbroken meadow, "level as a house-floor," but diversified by water-courses, "sloughs," and gentle acclivities—sometimes by considerable ravines and "barrens" or elevated "swales," thinly covered with timber, or brush, or both. But I will contemplate more especially a Northern farm, made up of hill and vale or glade, rocky ridge and skirting bog or other low land, with a wood-lot on

(39)

the rear or not far distant, and clumps or belts of
timber irregularly lining brook and ravine, or lurk-
ing in the angles and sinuosities of walls and wooden
fences, and a ragged, mossy orchard sheltered in some
quiet nook, or sprawling over some gravelly hill-side.
A brook, nearly dry in August, gurgles down the
hill-side or winds through the swamp; while fields,
moderately sloping here and nearly level there, interposed as they can be, have severally been devoted,
for a generation or more, alternately to Grain and
Grass—the latter largely preponderating. We will
suppose this farm to measure from 50 to 150 acres.

Now, the young man who has bought or inherited
this farm may be wholly and consciously unable to
enter upon any expensive system of improvement
for the next ten years—may fully realize that four or
five days of each week must meantime be given to
the growing or earning of present bread—yet he
should none the less study well the capacities and
adaptations of each acre, and mature a comprehen-
sive plan for the ultimate bringing of each field into
the best and most useful condition whereof it is susceptible, before he cuts a living tree or digs a solitary
drain. He is morally certain of doing something—
perhaps many things—that he will sadly wish un-
done, if he fails to study peculiarities and mature a
plan before he begins to improve or to fit his several
fields for profitable cultivation.

And the first selection to be made is that of a pasture, since I am compelled to use an old, familiar

name for what should be essentially a new thing.
This pasture should be as near the center of the farm
as may be, and convenient to the barns and barn-
yard that are to be.   It should have some shade, but
no very young trees; should be dry and rolling, with
an abundance of the purest living water.   The
smaller this pasture-lot may be, the better I shall like
it, provided you fence it very stoutly, connect it with
the barn-yard by a lane if they are not in close proximity, and firmly resolve that, outside of this lot, this
lane, this yard and the adjacent stable, your cattle
shall never be seen, unless on the road to market.
Very possibly, the day may come wherein you will
decide to dispense with pasturing altogether; but
that is, for the present, improbable. *One* pasture you
will have; if you live in the broad West, and purpose
to graze extensively, it will doubtless be a large one;
but permitting your stock to ramble in Spring and
Fall all over your own fields—(and perhaps your
neighbors' also)—in quest of their needful food, biting
off the tops of the finer young trees, trampling down
or breaking off some that are older, rubbing the
bark off of your growing fruit-trees, and doing damage that years will be required to repair, I most
vehemently protest against.

The one great error that misleads and corrupts
mankind is the presumption that *something may be
had for nothing.* The average farmer imagines that
whatever of flesh or of milk may accrue to him from
the food his cattle obtain by browsing over his fields

or through his woods, is so much clear gain—that they do the needful work, while he pockets the net proceeds. But the universe was framed on a plan which requires so much for so much; and this law will not submit to defiance or evasion. Under the unnatural, exceptional conditions which environ the lone squatter on a vast prairie, something may be made by turning cattle loose and letting them shift for themselves; but this is at best transitory, and at war with the exigencies of civilization. Whoever lives within sight of a school-house, or within hearing of a church-bell, is under the dominion of a law alike inexorable and beneficent—the law that requires each to pay for all he gets, and reap only where he has sown.

You can hardly have a pasture so small that it will not afford hospitality to weeds and prove a source of multiform infestations. The plants that should mature and be diffused will be kept down to the earth; those which should be warred upon and eradicated will flourish untouched, ripen their seed, and diffuse it far and wide. Thistles, White Daisy, and every plant that impedes tillage and diminishes crops, are nourished and diffused by means of pastures.

I hold, therefore, that the good farmer will run a mowing-machine over his pasture twice each Summer—say early in June, and then late in July—or, if his lot be too rough for this, will have it clipped at least once with a scythe. Cutting all manner of worthless if not noxious plants in the blossom, will benefit the soil which their seeding would tax; it will render the eradication of weeds from your tillage a far easier task; and' it will prevent your being a nuisance to your neighbors. I am confident that no one who has formed the habit of keeping down the weeds in his pasture will ever abandon it.

I think each pasture should have (though mine, as yet, has not) a rude shed or other shelter whereto the cattle may resort in case of storm or other inclemency. How much they shrink as well as suffer from one cold, pelting rain, few fully realize; but I am sure that "the merciful man" who (as the Scripture says) "is merciful to his beast," finds his humanity a good paying investment. I doubt that the rule would fail, even in Texas; but I am contemplating civilized husbandry, not the rude conditions of tropical semi-barbarism. If only by means of stakes and straw, give cattle a chance to keep dry and warm when they must otherwise shiver through a rainy, windy day and night on the cold, wet ground, and I am sure they will pay for it.

In confining a herd of cattle to such narrow limits, I do not intend that they shall be stinted to what grows there. On the contrary, I expect them to be fed on Winter Rye, on Cut Grass, on Sowed Corn, Sorghum, Stalks, Roots, etc., etc., as each shall be in season. With a good mower, it is a light hour's work before breakfast to cut and cart to a dozen or twenty head as much grass or corn as they will eat during the day. But let that point stand over for the present.

VII.

TREES—WOODLAND—FORESTS.

I am not at all sentimental—much less mawkish—regarding the destruction of trees. Descended from several generations of timber-cutters (for my paternal ancestors came to America in 1640), and myself engaged for three years in land-clearing, I realize that trees exist for use rather than for ornament, and have no more scruple as to cutting timber in a forest than as to cutting grass in a meadow. Utility is the reason and end of all vegetable growth—of a hickory's no less than a cornstalk's. I have always considered "Woodman, spare that tree," just about the most mawkish bit of badly versified prose in our language, and never could guess how it should touch the sensibilities of any one. Understand, then, that I urge the planting of trees mainly because I believe it will *pay*, and the preservation, improvement, and extension, of forests, for precisely that reason.

Yet I am not insensible to the beauty and grace lent by woods, and groves, and clumps or rows of trees, to the landscape they diversify. I feel the force of Emerson's averment, that "Beauty is its

(44)

own excuse for being," and know that a homestead embowered in, belted by, stately, graceful elms, maples, and evergreens, is really *worth* more, and will sell for more, than if it were naked field and meadow. I consider it one positive advantage (to balance many disadvantages) of our rocky, hilly, rugged Eastern country, that it will never, in all probability, be so denuded of forests as the rich, facile prairies and swales of the Great Valley may be. Our winds are less piercing, our tornadoes less destructive, than those of the Great West. I doubt whether there is another equal area of the earth's surface whereon so many kinds of valuable trees grow spontaneously and rapidly, defying eradication, as throughout New-England and on either slope of the Alleghenies ; and this profusion of timber and foliage may well atone for, or may be fairly weighed against, many deficiencies and drawbacks. The Yankee, who has been accustomed to see trees spring up spontaneously wherever they were not kept down by ax, or plow, or scythe, and to cross running water every half mile of a Summer day's journey, may well be made homesick, by two thousand miles of naked, dusty, wind-swept Plains, whereon he finds no water for fifty to a hundred miles, and knows it impossible to cut an ax-helve, much more an axle-tree, in the course of a wearying journey. No Eastern farmer ever realized the blessedness of abundant and excellent wood and water until he had wandered far from his boyhood's home.

No one may yet be able fully to explain the inter-dependence of these two blessings ; but the fact remains. All over " the Plains," there is evidence that trees grew and flourished where none are now found, and that springs and streams were then frequent and abiding where none now exist. A prominent citizen of Nevada, who explored southward from Austin to the Colorado, assured me that his party traveled for days in the bed of what had once been a considerable river, but in which it was evident that no water had flowed for years. And I have heard that, since the Mormons have planted trees over considerable sections of Utah, rains in Summer are no longer rare, and Salt Lake evinces, by a constant though moderate increase of her volume of waters, that the equilibrium of rain-fall with evaporation in the Great Basin has been fully restored—or rather, that the rain-fall is now taking the lead.

I have a firm faith that all the great deserts of the Temperate and Torrid Zones will yet be reclaimed by irrigation and tree-planting. The bill which Congress did not pass, nor really consider, whereby it was proposed, some years since, to give a section of the woodless Public Lands remote from settlement to every one who, in a separate township, would plant and cherish a quarter-section of choice forest-trees, ought to have been passed—with modifications, perhaps, but preserving the central idea. Had ten thousand quarter-sections, in so many different townships of the Plains, been thus planted to timber ten to twenty years ago, and protected from fire and devastation till now, the value of those Plains for settlement would have been nearly or quite doubled.

A capital mistake, it seems to me, is being made by some of the dairy farmers of our own State. One who has a hundred acres of good soil, whereof twenty or thirty are wooded, cuts off his timber entirely, calculating that the additional grass that he may grow in its stead will pay for all the coal he needs for fuel, so that he will make a net gain of the time he has hitherto devoted each Winter to cutting and hauling wood. He does not consider how much his soil will lose in Summer moisture, how his springs and runnels will be dried up, nor how the sweep of harsh winds will be intensified, by baring his hill-tops and ravines to sun and breeze so utterly. In my deliberate judgment, a farm of one hundred acres will yield *more* feed, with far greater uniformity of product from year to year, if twenty acres of its ridge-crests, ravine-sides, and rocky places, are thickly covered with timber, than if it be swept clean of trees and all devoted to grass. Hence, I insist that the farmer who sweeps off his wood and resolves to depend on coal for fuel, hoping to increase permanently the product of his dairy, makes a sad miscalculation.

Spain, Italy, and portions of France, are now suffering from the improvidence that devoured their forests, leaving the future to take care of itself. I presume the great empires of antiquity suffered from the same folly, though to a much greater extent. The

remains of now extinct races who formerly peopled
and tilled the central valleys of this continent, and
especially the Territory of Arizona, probably bear
witness to a similar recklessness, which is paralleled
by our fathers' and our own extermination of the
magnificent forests of White Pine which, barely a
century ago, covered so large a portion of the soil of
our Northern States. Vermont sold White Pine
abundantly to England through Canada within my
day : she is now supplying her own wants from Canada at a cost of not less than five times the price she
sold for ; and she will be paying still higher rates before the close of this century. I entreat our farmers
not to preserve every tree, good, bad, or indifferent,
that may happen to be growing on their lands—but,
outside of the limited districts wherein the primitive
forest must still be cut away in order that land may
be obtained for cultivation, to *plant and rear at least
two better trees for every one they may be impelled to
cut down.* How this may, in the average, be most
judiciously done, I will try to indicate in the succeeding chapter.

VIII.

GROWING TIMBER—TREE-PLANTING.

In my judgment, the proportion of a small farm that should be constantly devoted to trees (other than fruit) is not less than one-fourth; while, of farms exceeding one hundred acres in area, that proportion should be not less than one-third, and may often be profitably increased to one-half. I am thinking of such as are in good part superficially rugged and rocky, or sandy and sterile, such as New-England, eastern New-York, northern New-Jersey, with both slopes of the Alleghenies, as well as the western third of our continent, abound in. It may be that it is advisable to be content with a smaller proportion of timber in the Prairie States and the broad, fertile intervales which embosom most of our great rivers for at least a part of their course; but I doubt it. And there is scarcely a farm in the whole country, outside of the great primitive forests in which openings have but recently been made, in which *some* tree-planting is not urgently required.

"Too much land," you will hear assigned on every side as a reason for poor farming and meager crops.

3   (49)

Ask an average farmer in New-England, in Virginia, in Kentucky, or in Alabama, why the crops of his section are in the average no better, and the answer, three times in four, will be, " Our farmers have too much land"—that is, not too much absolutely, but too much relatively to their capital, stock, and general ability to till effectively. The habitual grower of poor crops will proffer this explanation quite as freely and frequently as his more thrifty neighbor. And what every one asserts must have a basis of truth.

Now, I do not mean to quarrel with the instinct which prompts my countrymen to buy and hold too much land. They feel, as I do, that land is still cheap almost anywhere in this country—cheap, if not in view of the income now derived from it, certainly in contemplation of the price it must soon command and the income it might, under better management, be made to yield. Under this conviction—or, if you please, impression—every one is intent on holding on to more land than he can profitably till, if not more than he can promptly pay for.

What I *do* object to is simply this—that thousands, who have more land than they have capital to work profitably, will persist in half-tilling many acres, instead of thoroughly farming one-half or one third so many, and getting the rest into wood so fast as may be. I am confident that two-thirds of all our farmers would improve their circumstances and increase their incomes by concentrating their efforts, their means, their fertilizers, upon half to two-thirds of the area they now skim and skin, and giving the residue back to timber-growing.

In my own hilly, rocky, often boggy, Westchester —probably within six of being the oldest Agricultural County in the Union—I am confident that ten thousand acres might to-morrow be given back to forest with profit to the owners and advantage to all its inhabitants. It is a fruit-growing, milk-producing, truck-farming county, closely adjoining the greatest city of the New World ; hence, one wherein land can be cultivated as profitably as almost anywhere else— yet I am satisfied that half its surface may be more advantageously devoted to timber than' to grass or tillage. Nay ; I doubt that one acre in a hundred of rocky land—that is, land ribbed or dotted with rocks that the bar or the rock-hook cannot lift from their beds, and which it will not as yet pay to blast —is now tilled to profit, or ever will be until it shall be found advisable to clear them utterly of stone breaking through or rising within two feet of the surface. The time will doubtless arrive in which many fields will pay for clearing of stone that would not to-day ; these, I urge, should be given up to wood now, and kept wooded until the hour shall have struck for ridding them of every impediment to the steady progress of both the surface and the subsoil plow.

Were all the rocky crests and rugged acclivities of this County bounteously wooded once more, and kept so for a generation, our floods would be less injuri-

ous, our springs unfailing, and our streams more constant and equable; our blasts would be less bitter,
and our gales less destructive to fruit; we should
have vastly more birds to delight us by their melody
and aid us in our not very successful war with devouring insects; we should grow peaches, cherries,
and other delicate fruits, which the violent caprices
of our seasons, the remorseless devastations of our
visible and invisible insect enemies, have all but an-
nihilated; and we should keep more cows and make
more milk on two-thirds of the land now devoted to
grass than we actually do from the whole of it. And
what is true of Westchester is measurably true of
every rural county in the Union.

I have said that I believe in cutting trees as well
as in growing them; I have not said, and do not
mean to say, that I believe in cutting everything
clean as you go. That was once proper in Westchester; it is still advisable in forest-covered regions,
where the sun must be let in before crops can be
grown; but, in nine cases out of ten, timber should
be thinned or culled out rather than cut off; and, for
every tree taken away, at least two should be planted
or set out.

We have pretty well outgrown the folly of letting
every apple-tree bear such fruit as it will; though in
the orchard of my father's little farm in Amherst,
N. H., whereon I was born, no tree had ever been
grafted when I bade adieu to it in 1820; and I presume none has been to this day. By this time, almost

every farmer realizes that he *can't afford* to grow little, gnarly, villainously sour or detestably bittersweet apples, when, by duly setting a graft at a cost of two dimes, he may make that identical tree yield Greenings or Pippins at least as bounteously. I presume the cumulative experience of fifty or sixty generations of apple-growers has ripened this conclusion. Why do they not infer readily and generally that growing indifferent timber where the best and most valued would grow as rapidly, is a stupid, costly blunder? It seems to me that whoever has attained the conviction that apple-trees should be grafted ought to know that it is wasteful to grow Red Oak, Beech, White Maple, and Alder, where White Oak, Hickory, Locust, and White Pine, might be grown with equal facility, in equal luxuriance, provided the right seeds were planted, and a little pains taken to keep down, for a year or two, the shoots spontaneously sent up by the wrong ones.

North of the Potomac, and east of the Ohio, and (I presume) in limited districts elsewhere, rocky, sterile woodlands, costing $2 to $50 per acre according to location, etc., are to-day the cheapest property to be bought in the United States. Even though nothing were done with them but keep out fire and cattle, and let the young trees grow as they will, money can be more profitably and safely invested in lands covered by young timber than in anything else. The parent, who would invest a few thousands for the benefit of children or grandchildren still young, may buy woodlands which will be worth twenty times their present cost within the next twenty years. But better even than this would it be to buy up rocky, craggy, naked hill-sides and eminences which have been pastured to death, and, shutting out cattle inflexibly, scratch these over with plow, mattock, hoe, or pick, as circumstances shall dictate, plant them thickly with Chestnut, Walnut, Hickory, White Oak, and the seeds of Locust and White Pine. I say Locust, though not yet certain that this tree must not be started in garden or nursery-beds and transplanted when two or three years old, so puny and feeble is it at the outset, and so likely to be smothered under leaves or killed out by its more favored neighbors. I have experiments in progress not yet matured, which may shed light on this point before I finish these essays.

Plant thickly, and of diverse kinds, so as to cover the ground promptly and choke out weeds and shrubs, with full purpose to thin and prune as circumstances shall dictate.

Many farmers are averse to planting timber, because (they think) nothing can be realized therefrom for the next twenty or thirty years, which is as long as they expect to live. But this is a grave miscalculation. Let us suppose a rocky, hilly pasture-lot of ten or twenty acres rudely scratched over as I have suggested, and thickly seeded with hickory nuts and white oak acorns only: within five years, it will yield abundantly of hoop-poles, though the better, more promising half be left to mature, as they should be; two years later, another and larger crop of hoop-poles may be cut, still sparing the best; and thenceforth a valuable crop of timber may be taken from that land; for, if cut at the proper season, at least two thrifty sprouts will start from every stump; and so that wood will yield a clear income each year while its best trees are steadily growing and maturing. I do not advise restriction to those two species of timber; but I insist that a young plantation of forest-trees may and should yield a clear income in every year after its fourth.

As to the Far West—the Plains, the Parks, and the Great Basin—there is more money to be made by dotting them with groves of choice timber than by working the richest veins of the adjacent mountains. Whoever will promptly start, near a present or prospective railroad, forty acres of choice trees—Hickory, White Oak, Locust, Chestnut, and White Pine —within a circuit of three hundred miles from Denver, on land which he has made or is making provision to irrigate—may begin to sell trees therefrom two years hence, and persist in selling annually henceforth for a century—at first, for transplanting; very soon, for a variety of uses in addition to that.

—But this paper grows too long, and I must postpone to the next my more especial suggestions to young farmers with regard to tree-planting.

IX.

WHOEVER has recently bought, inherited, or otherwise become the owner of a farm, has usually found some part or parts of it devoted to wood; and this, if not in excess, he will mainly preserve, while he studies and plans with a view to the ultimate devotion to timber of just those portions of his land that are best adapted to that use. In locating that timber, I would have him consider these suggestions:

I. Land wisely planted with trees, and fenced so far as need be to keep out cattle, costs nothing. Whatever else you grow involves labor and expenditure; trees grow of their own accord. You may neglect them utterly—may wander over the earth and be absent for ten or twenty years, while your fences decay and your fields are overcropped to exhaustion; even your meadows may be run out by late mowing and close feeding at both ends of the season, till a dozen acres will hardly subsist a span of horses and a cow; but your woods need only to be let alone to insure that their value shall have decidedly increased during your absence. They will

(56)

richly reward labor and care in thinning, trimming, and transplanting — you may profitably employ in them any time that you can spare them—but they will do very well if simply let alone. And, unlike any other product with which I am acquainted, you may take crop after crop of wood from the same lot, and the soil will be richer and more productive after the last than it was before the first. Whether wholly because their roots permeate and break up the soil during their life and enrich it in their decay, or for diverse reasons, it is certainly true that land—and especially *poor* land—is enriched by growing upon it a crop of almost any timber, the evergreens possibly excepted. So, should you ever have land that you cannot till to profit, whether because it is too poor, or because you have a sufficiency that is better, you should at once devote it to wood.

II. Your springs and streams will be rendered more equable and enduring by increasing the area and the luxuriance of your timber. They may have become scanty and capricious under a policy of reckless, wholesale destruction of trees ; they will be reenforced and reinvigorated by doubling the area of your woods, while quadrupling the number, and increasing the average size, of your trees.

III. All ravines and steep hill-sides should be devoted to trees. Every acre too rocky to be thoroughly cleared of stone and plowed should be set apart for tree-growing. Wherever the soil will be gullied or washed away by violent rains if under tillage, it should be excluded from cultivation and given up to trees. Men often doubt the profit of heavy manuring; and well they may, if three-fourths of the fertilizers applied are soaked out and swept away by flooding rains or sudden thaws and floated off to some distant sea or bay; but let all that is applied to the soil only remain there till it is carted away in crops, and it will hardly be possible to manure too highly for profit.

IV. Trees, especially evergreens, may be so disposed as to modify agreeably the average temperature of your farm, or at least of the most important parts of it. When I bought my place—or rather the first installment of it—the best spot I could select for a garden lay at the foot of a hill which half surrounded it on the south and east, leaving it exposed to the full sweep of north and north-west winds; so that, though the soil was gravelly and warm, my garden was likely to be cold and backward. To remedy this, I planted four rows of evergreens (Balsam Fir, Pine, Red Cedar, and Hemlock), along a low ridge bounding it on the north, following an inward curve of the ridge at its west end; and those evergreens have in sixteen years grown into very considerable trees, forming a shady, cleanly, inviting bower, or sylvan retreat, daintily carpeted with the fallen leaves of the overhanging firs. I judge that the average temperature of the soil for some yards southward of this wind-break is at least five degrees higher, throughout the growing season, than it formerly was or would now be if these evergreens were swept away; while the aspect of the place is agreeably diversified, and even beautified, by their appearance. I believe it would sell for some hundreds of dollars more with than without that thrifty, growing clump of evergreens.

V. I have already urged, though not strongly enough, that crops, as well as springs, will be improved by keeping the crests of ridges thickly wooded, thus depositing moisture in Winter and Spring, to be slowly yielded to the adjacent slopes during the heat and drouth of Summer. I firmly believe that the slopes of a hill whose crest is heavily wooded will yield larger average crops than slope and crest together would do if both were bare of trees.

VI. The banks of considerable streams, ponds, etc., may often be so planted with trees that these will shade more water than land, to the comfort and satisfaction of the fish, and the protection of those banks from abrasion by floods and rapid currents. Sycamore, Elm, and Willow, do well here; if choice Grape-Vines are set beside and allowed to run over some of them, the effect is good, and the grapes acceptable to man and bird.

VII. Never forget that a good tree grows as thriftily and surely as a poor one. Many a farmer has to-day ten to forty acres of indifferent cord-wood where he might, at a very slight cost, have had instead an equal quantity of choice timber, worth ten times

as much.   Hickory, Chestnut, and Walnut, while
they yield nuts that can be eaten or sold, are worth
far more as timber than an equal bulk of ·Beech,
Birch, Hemlock, or Red Oak.   Chestnut has more
than doubled in value within the last few years,
mainly because it has been found excellent for the
inside wood-work of dwellings.   Locust also seems
to be increasing in value.   Ten acres of large, thrifty
Locust near this City would now buy a pretty good
farm ; as I presume it would, if located near any of
our great cities.

VIII. Where several good varieties of Timber are
grown together, some insect or atmospheric trouble
may blast one of them, yet leave the residue alive
and hearty.   And, if all continue thrifty, some may
be cut out and sold, leaving others more room to
grow and rapidly attain a vigorous maturity.

IX. Wherever timber has become scarce and valuable, a wood-lot should be thinned out, nevermore
cleared off, unless it is to be devoted to a different use.
It seems to me that destroying a forest because we
want timber is like smothering a hive of bees because
we want honey.

X. Timber should be cut with intelligent reference
to the future.   Locust and other valuable trees that
it is desirable should throw up shoots from the
stump, and rapidly reproduce their kind, should be
cut in March or April ; while trees that you want
to exterminate should be cut in August, so that they
may *not* sprout.   There may be exceptions to this

rule; but I do not happen to recollect any. Evergreens do not sprout; and I think these should be cut in Winter—at all events, not in Spring, when full of sap and thus prone to rapid decay.

XI. Your plantation will furnish pleasant and profitable employment at almost any season. I doubt that any one in this country has ever yet bestowed so much labor and care on a young forest as it will amply reward. Sow your seeds thickly; begin to thin the young trees when they are a foot high, and to trim them so soon as they are three feet, and you may have thousands thriving on a fertile acre, and pushing their growth upward with a rapidity and to an altitude outrunning all preconception.

XII. Springs and streams will soon appear where none have appeared and endured for generations, when we shall have rëclothed the nakedness of the Plains with adequate forests. Rains will become moderately frequent where they are now rare, and confined to the season when they are of least use to the husbandman.

I may have more to say of trees by-and-by, but rest here for the present. The importance of the topic can hardly be overrated.

X.

DRAINING—MY OWN.

My farm is in the township of Newcastle West-chester County, N. Y., 35 miles from our City Hall, and a little eastward of the hamlet known as Chappaqua, called into existence by a station on the Harlem Railroad. It embraces the south-easterly half of the marsh which the railroad here traverses from south to north—my part measuring some fifteen acres, with five acres more of slightly elevated dry land between it and the foot of the rather rugged hill which rises thence on the east and on the south, and of which I now own some fifty acres, lying wholly eastward of my low land, and in good part covered with forest. Of this, I bought more than half in 1853, and the residue in bits from time to time as I could afford it. The average cost was between $130 and $140 per acre: one small and poor old cottage being the only building I found on the tract, which consisted of the ragged edges of two adjacent farms, between the western portions of which mine is now interposed, while they still adjoin each other beyond the north and south road, half a mile

from the railroad, on which their buildings are located
and which forms my eastern boundary. My stony,
gravelly upland mainly slopes to the west; but two
acres on my east line incline toward the road which
bounds me in that direction, while two more on my
south-east corner descend to the little brook which,
entering at that corner, keeps irregularly near my
south line, until it emerges, swelled by a smaller runnel that enters my lowland from the north and traverses it to meet and pass off with the larger brooklet aforesaid. I have done some draining, to no great
purpose, on the more level portions of my upland;
but my lowland has challenged my best efforts in
this line, and I shall here explain them, for the en-
couragement and possible guidance of novices in
draining. Let me speak first of

My Difficulties.—This marsh or bog consisted,
when I first grappled with it, of some thirty acres,
whereof I then owned less than a third. To drain it
to advantage, one person should own it all, or the
different owners should coöperate; but I had to go
it alone, with no other aid than a freely accorded
privilege of straightening as well as deepening the
brook which wound its way through the dryer meadow just below me, forming here the boundary of
two adjacent farms. I spent $100 on this job,
which is still imperfect; but the first decided fall in
the stream occurs nearly a mile below me; and you
tire easily of doing at your own cost work which
benefits several others as much as yourself. My

drainage will never be perfect till this brook, with that far larger one in which it is merged sixty rods below me, shall have been sunk three or four feet, at a further expense of at least $500.

This bog or swamp, when I first bought into it, was mainly dedicated to the use of frogs, muskrats and snapping-turtles. A few small water-elms and soft maples grew upon it, with swamp alder partly fringing the western base of the hill east of it, where the rocks which had, through thousands of years, rolled from the hill, thickly covered the surface, with springs bubbling up around and among them. Decaying stumps and imbedded fragments of trees argued that timber formerly covered this marsh as well as the encircling hills. A tall, dense growth of blackberry briers, thoroughwort, and all manner of marsh-weeds and grasses, covered the center of the swamp each Summer; but my original portion of it, being too wet for these, was mainly addicted to hassocks or tussocks of wiry, worthless grass; their matted roots rising in hard bunches a few inches above the soft, bare, encircling mud. The bog ranged in depth from a few inches to five or six feet, and was composed of black, peaty, vegetable mold, diversified by occasional streaks of clay or sand, all resting on a substratum of hard, coarse gravel, out of which two or three springs bubbled up, in addition to the half a dozen which poured in from the east, and a tiny rivulet which (except in a very dry, hot time) added the tribute of three or four more, which sprang from the base of a higher shelf of the hill near the middle of what is now my farm. Add to these that the brook which brawled and foamed down my hill-side near my south line as aforesaid, had brought along an immensity of pebbles and gravel of which it had mainly formed my five acres of dryer lowland, had thus built up a pretty swale, whereon it had the bad habit of filling up one channel, and then cutting another, more devious and eccentric, if possible, than any of its predecessors— and you have some idea of the obstacles I encountered and resolved to overcome. One of my first substantial improvements was the cutting of a straight channel for this current and, by walling it with large stones, compelling the brook to respect necessary limitations. It was not my fault that some of those stones were set nearly upright, so as to veneer the brook rather than thoroughly constrain it: hence, some of the stones, undermined by strong currents, were pitched forward into the brook by high Spring freshets, so as to require resetting more carefully. This was a mistake, but not one of

My Blunders.—These, the natural results of inexperience and haste, were very grave. Not only had I had no real experience in draining when I began, but I could hire no foreman who knew much more of it than I did. I ought to have begun by securing an ample and sure fall where the water left my land, and next cut down the brooklet or open ditch into which I intended to drain to the lowest practicable point—so low, at least, that no drain running into it should ever be troubled with back-water. Nothing can be more useless than a drain in which water stagnates, choking it with mud. Then I should have bought hundreds of Hemlock or other cheap boards, slit them to a width of four or five inches, and, having opened the needed drains, laid these in the bottom and the tile thereupon, taking care to *break joint*, by covering the meeting ends of two boards with the middle of a tile. Laying tile in the soft mud of a bog, with nothing beneath to prevent their sinking, is simply throwing away labor and money. I cannot wonder that tile-draining seems to many a humbug, seeing that so many tile are laid so that they can never do any good.

Having, by successive purchases, become owner of fully half of this swamp, and by repeated blunders discovered that making stone drains in a bog, while it is a capital mode of getting rid of the stone, is no way at all to dry the soil, I closed my series of experiments two years since by carefully rëlaying my generally useless tile on good strips of board, sinking them just as deep as I could persuade the water to run off freely, and, instead of allowing them to discharge into a brooklet or open ditch, connecting each with a covered main of four to six-inch tile; these mains discharging into the running brook which drains all my farm and three or four of those above it just where it runs swiftly off from my land. If a thaw or heavy rain swells the brook (as it sometimes will) so that it rises above my outlet aforesaid, the strong current formed by the concentration of the clear contents of so many drains will not allow the muddy water of the brook to back into it so many as three feet at most; and any mud or sediment that may be deposited there will be swept out clean whenever the brook shall have fallen to the drainage level. For this and similar excellent devices, I am indebted to the capital engineering and thorough execution of Messrs. Chickering & Gall, whose work on my place has seldom required mending, and never called for reconstruction.

My Success.—I judge that there are not many tracts more difficult to drain than mine was, considering all the circumstances, except those which are frequently flowed by tides or the waters of some lake or river. Had I owned the entire swamp, or had there been a fall in the brook just below me, had I had any prior experience in draining, or had others equally interested coöperated in the good work, my task would have been comparatively light. As it was, I made mistakes which increased the cost and postponed the success of my efforts; but this is at length complete. I had seven acres of Indian Corn, one of Corn Fodder, two of Oats, and seven or eight acres of Grass, on my lowland in 1869; and, though the Spring months were quite rainy, and the latter part of Summer rather dry, my crops were all good. I did not see better in Westchester County; and I shall be quite content with as good hereafter. Of my seven hundred bushels of Corn (ears,) I judge that two-thirds would be accounted fit for seed anywhere; my Grass was cut twice, and yielded one large crop and another heavier than the average first crop throughout our State. My drainage will require some care henceforth; but the fifteen acres I have reclaimed from utter uselessness and obstructions are decidedly the best part of my farm. Uplands may be exhausted; these never can be.

The experience of another season (1870) of protracted drouth has fully justified my most sanguine expectations. I had this year four acres of Corn, and as many of Oats, on my swamp, with the residue in Grass; and they were all good. I estimate my first Hay-crop at over two and a half tuns per acre, while the rowen or aftermath barely exceeded half a tun per acre, because of the severity of the drouth, which began in July and lasted till October. My Oats were good, but not remarkably so; and I had 810 bushels of ears of sound, ripe Corn from four acres of drained swamp and two and a half of upland. I estimate my upland Corn at seventy (shelled) bushels, and my lowland at fifty-five (shelled) bushels per acre. Others, doubtless, had more, despite the unpropitious season; but my crop was a fair one, and I am content with it. My upland Corn was heavily manured; my lowland but moderately. There are many to tell you how much I lose by my farming; I only say that, as yet, no one else has lost a farthing by it, and I do not complain.

XI.

DRAINING GENERALLY.

HAVING narrated my own experience in draining with entire unreserve, I here submit the general conclusions to which it has led me ·

I. While I doubt that there is *any* land above water that would not be improved by a good system of underdrains, I am sure that there is a great deal that could not at present be drained to profit. Forests, hill-side pastures, and most dry gravelly or sandy tracts, I place in this category. Perhaps one-third of New-England, half of the Middle States, and three-fourths of the Mississippi Valley, may ultimately be drained with profit.

II. *All* swamp lands without exception, nearly all clay soils, and a majority of the flat or gently rolling lands of this country, must eventually be drained, if they are to be tilled with the best results. I doubt that there is a garden on earth that would not be (unless it already had been) improved by thorough underdraining.

III. The uses of underdrains are many and diverse. To carry off surplus water, though the most obvious, stands by no means alone. 1. Underdrained land may be plowed and sowed considerably earlier in Spring than undrained soil of like quality. 2. Drained fields lose far less than others of their fertility by washing. 3. They are not so liable to be gullied by sudden thaws or flooding rains. 4. Where a field has been deeply subsoiled, I am confident that it will remain mellow and permeable by roots longer than if undrained. 5. Less water being evaporated from drained than from undrained land, the soil will be warmer throughout the growing season; hence, the crop will be heavier, and will mature earlier. 6. Being more porous and less compact, I think the soil of a drained field retains more moisture in a season of drouth, and its growing plants suffer less therefrom, than if it were undrained. In short, I thoroughly believe in underdraining.

IV. Yet I advise no man to run into debt for draining, as I can imagine a mortgage on a farm so heavy and pressing as to be even a greater nuisance than stagnant water in its soil. Labor and tile are dear with us; I do not expect that either will ever be so cheap here as in England or Belgium. What I *would* have each farmer in moderate circumstances do is to *drain his wettest field* next Fall—that is, after finishing his haying and before cutting up his corn—taking care to secure abundant fall to carry off the water in time of flood, and doing his work thoroughly. Having done this, let him subsoil deeply, fertilize amply, till carefully, and watch the result.

I think it will soon satisfy him that such draining pays.

V. I do not insist on tile as making the only good drain; but I have had no success with any other. The use of stone, in my opinion, is only justified where the field to be drained abounds in them and no other use can be made of them. To make a good drain with ordinary boulders or cobble-stones requires twice the excavation and involves twice the labor necessarily expended on tile-draining; and it is neither so effective nor so durable. Earth will be carried by water into a stone drain; rats and other vermin will burrow in it and dig (or enlarge) holes thence to the surface; in short, it is not the thing. Better drain with stone where they are a nuisance than not at all; but I predict that you will dig them up after giving them a fair trial and replace them with tile. In a wooded country, where tile were scarce and dear, I should try draining with slabs or cheap boards dressed to a uniform width of six or eight inches, and laid in a ditch dug with banks inclined or sloped to the bottom, so as to form a sort of V; the lower edge of the two side-slabs coming together at the bottom, and a third being laid widely across their upper edges, so as to form a perfect cap or cover. In firm, hard soil, this would prove an efficient drain, and, if well made, would last twenty years. Uniformity of temperature and of moisture would keep the slabs tolerably sound for at least so long; and, if the top of this drain were two feet below the surface, no plowing or trampling over it would harm it.

VI. As to draining by what is called a Mole Plow, which simply makes a waterway through the subsoil at a depth of three feet or thereabout, I have no acquaintance with it but by hearsay. It seems to me morally impossible that drains so made should not be lower at some points than at others, so as to retain their fill of water instead of carrying it rapidly off; and I am sure that plowing, or even carting heavy loads over them, must gradually choke and destroy them. Yet this kind of draining is comparatively so cheap, and may, with a strong team, be effected so rapidly, that I can account for its popularity, especially in prairie regions. Where the subsoil is rocky, it is impracticable; where it is hard-pan, it must be very difficult; where it is loose sand, it cannot endure; but in clays or heavy loams, it may, for a few years, render excellent service. I wish the heavy clays of Vermont, more especially of the Champlain basin, were well furrowed or pierced by even such drains; for I am confident that they would temporarily improve both soil and crop; and, if they soon gave out, they would probably be replaced by others more durable.

—I shall not attempt to give instructions in drain-making; but I urge every novice in the art to procure Waring's or some other work on the subject and study it carefully: then, if he can obtain at a fair price the services of an experienced drainer, hire him to supervise the work. One point only do I insist on —that is, draining into a main rather than an open ditch or brook; for it is difficult in this or any harsher climate to prevent the crumbling of your outlet tile by frost. Below the Potomac or the Arkansas, this may not be apprehended; and there it may be best to have your drains separately discharge from a roadside bank or into an open ditch, as they will thus inhale more air, and so help (in Summer) to warm and moisten the soil above them; but in our climate I believe it better to let your drains discharge into a covered main or mains as aforesaid, than into an open ditch or brook.

Tile and labor are dear with us; I presume labor will remain so. But, in our old States, there are often laborers lacking employment in November and the Winter months; and it is the wisest and truest charity to proffer them pay for work. Some will reject it unless the price be exorbitant; but there are scores of the deserving poor in almost every rural county, who would rather earn a dollar per day than hang around the grog-shops waiting for Spring. Get your tiles when you can, or do not get them at all, but let it be widely known that you have work for those who will do it for the wages you can afford, and you will soon have somebody to earn your money. Having staked out your drains, set these to work at digging them, even though you should not be able to tile them for a year. Cut your outlet deep, and your land will profit by a year of open drains.

4

XII.

IRRIGATION — MEANS AND ENDS.

WHILE few can have failed to realize the important part played by Water in the economy of vegetation, I judge that the question—" How can I secure to my growing plants a sufficiency of moisture at all times?"—has not always presented itself to the farmer's mind as demanding of him a practical solution. To rid his soil and keep it free of superfluous, but especially of stagnant water, he may or may not accept as a necessity; but that, having provided for draining away whatever is excessive, he should turn a short corner and begin at once to provide that water shall be supplied to his fields and plants whenever they may need it, he is often slow to apprehend. Yet this provision is but the counterpart and complement of the other.

I had sped across Europe to Venice, and noted with interest the admirable, effective irrigation of the great plain of Lombardy, before I could call any land my own. I saw there a region perhaps thirty miles wide by one hundred and fifty along the east bank of the Po, rising very gently thence to the foot of the

(74)

Austrian Alps, which Providence seems to have specially adapted to be improved by irrigation. The torrents of melted snow which in Spring leap and foam adown the southern face of the Alps, bringing with them the finer particles of soil, are suddenly arrested and form lakes (Garda, Maggiore, Como, etc.) just as they emerge upon the plain. These lakes, slowly rising, often overflow their banks, with those of the small rivers that bear their waters westward to the Po; and this overflow was a natural source of abiding fertility. To dam these outlets, and thus control their currents, was a very simple and obvious device of long ago, and was probably begun by a very few individuals (if by more than one), whose success incited emulation, until the present extensive and costly system of irrigating dams and canals was gradually developed. When I traversed Lombardy in July, 1851, the beds of streams naturally as large as the Pemigewasset, Battenkill, Canada Creek, or Humboldt, were utterly dry; the water which would naturally have flowed therein being wholly transferred to an irrigating canal (or to canals) often two or three miles distant. The reservoirs thus created were filled in Spring, when the streams were fullest and their water richest, and gradually drawn upon throughout the later growing season to cover the carefully leveled and graded fields on either side to the depth of an inch or two at a time. If any failed to be soon absorbed by the soil, it was drawn off as here superfluous, and added to the current employed to moisten and fertilize the field next below it; and so field after field was refreshed and enriched, to the husbandman's satisfaction and profit. It may be that the rich glades of English Lancashire bear heavier average crops; but those of Lombardy are rarely excelled on the globe.

Why should not our Atlantic slope have its Lombardy? Utah, Nevada, and California, exhibit raw, crude suggestions of such a system; but why should the irrigation of the New World be confined to regions where it is indispensable, when that of the Old is not? I know no good reason whatever. for leaving an American field unirrigated where water to flow it at will can be had at a moderate cost.

When I first bought land (in 1853) I fully purposed to provide for irrigating my nearly level acres at will, and I constructed two dams across my upland stream with that view; but they were so badly planned that they went off in the flood caused by a tremendous rain the next Spring; and, though I rebuilt one of them, I submitted to a miscalculation which provided for taking the water, by means of a syphon, out of the pond at the top and over the bank that rose fifteen or twenty feet above the surface of the water. Of course, air would work into the pipe after it had carried a stream unexceptionably for two or three days, and then the water would run no longer. Had I taken it from the bottom of the pond through my dam, it would have run forever, (or so long as there was water covering its inlet in the pond;) but bad engineering flung me; and I have never since had the heart (or the means) to revise and correct its errors.

My next attempt was on a much humbler scale, and I engineered it myself. Toward the north end of my farm, the hill-side which rises east of my low-land is broken by a swale or terrace, which gives me three or four acres of tolerably level upland, along the upper edge of which five or six springs, which never wholly fail, burst from the rocks above and unite to form a petty runnel, which dries up in very hot or dry weather, but which usually preserved a tiny stream to be lost in the swamp below. North of the gully cut down the lower hill-side by this streamlet, the hill-side of some three acres is quite steep, still partially wooded, and wholly devoted to pasturage. Making a petty dam across this runnel at the top of the lower acclivity, I turned the stream aside, so that it should henceforth run along the crest of this lower hill, falling off gradually so as to secure a free current, and losing its contents at intervals through variable depressions in its lower bank. Dam and artificial water-course together cost me $90, which was about twice what it should have been. That rude and petty contrivance has now been ten years in operation, and may have cost $5 per annum for oversight and repairs. Its effect has been to double the grass grown on the two acres it constantly irrigates, for which I paid $280, or more than thrice the cost of my irrigation. But more: my hill-side, while

it was well grassed in Spring, always gave out directly after the first dry, or hot week; so that, when I
most needed feed, it afforded none; its herbage being
parched up and dead, and thus remaining till refreshed by generous rains. I judge, therefore, that my
irrigation has *more* than doubled the product of those
two acres, and that these are likely to lose nothing in
yield or value so long as that petty irrigating ditch
shall be maintained.

I know this is small business. But suppose each
of the hundred thousand New-England farms, whereof five to ten acres might be thus irrigated at a cost
not exceeding $100 per farm, had been similarly
prepared to flow those acres last Spring and early
Summer, with an average increase therefrom of
barely one tun of Hay (or its equivalent in pasturage)
per acre. The 500,000 tuns of Hay thus realized
would have saved 200,000 head of cattle from being
sent to the butcher while too thin for good beef,
while every one of them was required for further
use, and will have to be replaced at a heavy cost.
Shall not these things be considered? Shall not all
who can do so at moderate cost resolve to test on
their own farms the advantages and benefits that
may be secured by Irrigation?

XIII.

I HAVE given an account of my poor, little experiment in Irrigation, because it is one which almost every farmer can imitate and improve upon, however narrow his domain and slender his fortune. I presume there are Half a Million homesteads in the United States which have natural facilities for Irrigation at least equal to mine; many of them far greater. Along either slope of the Alleghenies, throughout a district at least a thousand miles long by three hundred wide, nearly every farm might be at least partially irrigated by means of a dam costing from twenty-five to one hundred dollars; so might at least half the farms in New-England and our own State. On the prairies, the plans must be different, and the expense probably greater, but the results obtained would bounteously reward the outlay. I shall not see the day, but there are those now living who *will* see it, when Artesian wells will be dug at points where many acres may be flowed from a gentle swell in the midst of a vast plain, or at the head of a fertile valley, expressly, or at least mainly, that its waters

(79)

may be led across that plain, adown that valley, in irrigating streams and ditches until they have been wholly drank up by the soil. I have seen single wells in California that might be made to irrigate sufficiently hundreds of acres, by the aid of a reservoir into which their waters could be discharged when the soil did not require them, and there retained until the thirsty earth demanded them.

An old and successful farmer in my neighborhood affirms that Water is the cheapest and best fertilizer ever applied to the soil. If this were understood to mean that no other is needed or can be profitably applied, it would be erroneous. Still, I think it clearly true that the annual product of most farms can be increased, and the danger of failure averted, more cheaply by the skillful application of water than by that of any other fertilizer whatever, Plaster (Gypsum) possibly excepted.

I took a run through Virginia last Summer, not far from the 1st of August. That State was then suffering intensely from drouth, as she continued to do for some weeks thereafter. I am quite sure that I saw on her thirsty plains and hillsides not less than three hundred thousand acres planted with Indian Corn, whereof the average product could not exceed ten bushels per acre, while most of it would fall far below that yield, and there were thousands of acres that would not produce one sound ear! Every one deplored the failure, correctly attributing it to the prevailing drouth. And yet, I passed hundreds if not thousands of places where a very moderate outlay would have sufficed to dam a stream or brooklet issuing from between two spurs of the Blue Ridge, or the Alleghenies, so that a refreshing current of the copious and fertilizing floods of Winter and Spring, warmed by the fervid suns of June and July, could have been led over broad fields lying below, so as to vanquish drouth and insure generous harvests. Nay; I feel confident that I could in many places have constructed rude works in a week, after that drouth began to be felt, that would have saved and made the Corn on at least a portion of the planted acres through which the now shrunken brooks danced and laughed idly down to the larger streams in the wider and equally thirsty valleys. Of course, I know that this would have been imperfect irrigation—a mere stop-gap— that the cold spring-water of a parched Summer cannot fertilize as the hill-wash of Winter and Spring, if thriftily garnered and warmed through and through for sultry weeks, would do; yet I believe that very many farmers might, even then, have secured partial crops by such irrigation as was still possible, had they, even at the eleventh hour, done their best to retrieve the errors of the past.

. For the present, I would only counsel every farmer to give his land a careful scrutiny with a view to irrigation in the future. No one is obliged to do any faster than his means will justify; and yet it may be well to have a clear comprehension of all that may ultimately be done to profit, even though much of

4*

it must long remain unattempted. In many cases, a stream may be dammed for the power which it will afford for two or three months of each year, if it shall appear that this use is quite consistent with its employment to irrigation, when the former alone would not justify the requisite outlay. It is by thus making one expense subserve two quite independent but not inconsistent purposes that success is attained in other pursuits ; and so it may be in farming.

As yet, each farmer must study his own resources with intent to make the most of them. If a manageable stream crosses or issues from his land, he must measure its fall thereon, study the lay of the land, and determine whether he can or cannot, at a tolerable cost, make that stream available in the irrigation of at least a portion of his growing crops when they shall need water and the skies decline to supply it. On many, I think on most, farms situated among hills, or upon the slopes of mountains, something may be done in this way—done at once, and with immediate profit. But this is rudimentary, partial, fragmentary, when compared with the irrigation which yet shall be. I am confident that there are points on the Carson, the Humboldt, the Weber, the South Platte, the Cache-le-Poudre, and many less noted streams which thrid the central plateau of our continent, where an expenditure of $10,000 to $50,000 may be judiciously made in a dam, locks and canals, for the purposes of irrigation and milling combined, with a moral certainty of realizing fifty per cent. annually on the outlay, with a steady increase in the value of the property. If my eye did not deceive me, there is one point on the Carson where a dam that need not cost $50,000 would irrigate one hundred square miles of rich plain which, when I saw it eleven years ago, grew nought but the worthless shrubs of the desert, simply because nothing else could endure the intense, abiding drouth of each Nevada Summer. Such palpable invitations to thrift cannot remain forever unimproved.

In regions like this, where Summer rains are the rule rather than the exception, the need of irrigation is not so palpable, since we do or may secure decent average crops in its absence. Yet there is no farm in our country that would not yield considerably more grain and more grass, more fruit and more vegetables, if its owner had water at command which he could apply at pleasure and to any extent he should deem requisite. Most men, thus empowered, would at first irrigate too often and too copiously; but experience would soon temper their zeal, and teach them

"The precious art of Not too much;"

and they would thenceforth be careful to give their soil drink yet, not drown it.

Whoever lives beyond the close of this century, and shall then traverse our prairie States, will see them whitened at intervals by the broad sails of windmills erected over wells, whence every gale or breeze will be employed in pumping water into the ponds or reservoirs so located that water may be drawn therefrom at will and diffused in gentle streamlets over the surrounding fields to invigorate and impel their growing crops. And, when all has been done that this paper faintly foreshadows, our people will have barely indicated, not by any means exhausted, the beneficent possibilities of irrigation.

The difficulty is in making a beginning. Too many farmers would fain conceal a poverty of thought behind an affectation of dislike or contempt for novelties. " Humbug ! " is their stereotyped comment on every suggestion that they might wisely and profitably do something otherwise than as their grandfathers did. They assume that those respected ancestors did very well without Irrigation ; wherefore, it cannot now be essential. But the circumstances have materially changed. The disappearance of the dense, high woods that formerly almost or quite surrounded each farm has given a sweep to the heated, parching winds of Summer, to which our ancestors were strangers. Our springs, our streams, do not hold out as they once did. Our Summer drouths are longer and fiercer. Even though our grandfathers did not, we *do* need and may profit by Irrigation.

XIV.

PLOWING—DEEP OR SHALLOW.

Rules absolutely without exception are rare ; and they who imagine that I insist on plowing all lands deeply are wrong for I hold that much land should never be plowed at all. In fact, I have seen in my life nearly as large an area that ought not as I have that ought to be plowed, by which I mean that half the land I have seen may serve mankind better if devoted to timber than if subjected to tillage. I personally know farmers who would thrive far better if they tilled but half the area they do, bestowing on this all the labor and fertilizers they spread over the whole, even though they threw the residue into common and left it there. I judge that a majority of our farmers could increase the recompense of their toil by cultivating fewer acres than they now do.

Nor do I deny that there are soils which it is not advisable to plow deeply. Prof. Mapes told me he had seen a tract in West Jersey whereof the soil was but eight inches deep, resting on a stratum of copperas (sulphate of iron,) which, being upturned by the plow and mingled with the soil, poisoned the

crops planted thereon.   And I saw, last Summer, on
the intervale of New River, in the western part of
Old Virginia, many acres of Corn which were thrifty
and luxuriant in spite of shallow plowing and intense drouth, because the rich, black loam which had
there been deposited by semi-annual inundations,
until its depth ranged from two to twenty feet, was so
inviting and permeable that the corn-roots ran *below*
the bottom of the furrow about as readily as above
that line.   I do not doubt that there are many millions of acres of such land that would produce tolerably, and sometimes bounteously, though simply
scratched over by a brush harrow and never plowed
at all.   In the infancy of our race, when there were
few mouths to fill and when farming implements
were very rude and ineffective, cultivation was all
but confined to these facile strips and patches, so
that the utility, the need, of deep tillage was not apparent.   And yet, we know the crops often failed
utterly in those days, plunging whole nations into
the miseries of famine.

The primitive plow was a forked stick or tree-top,
whereof one prong formed the coulter, the other and
longer the beam; and he who first sharpened the
coulter-prong with a stone hatchet was the Whitney
or McCormick of his day.   The plow in common use
to-day in Spain or Turkey is an improvement on
this, for it has an iron point; still, it is a miserable
tool.   When, at five years old, I first rode the horse
which drew my father's plow in furrowing for or culti-

vating his corn, it had an iron coulter and an iron share; but it was mainly composed of wood. In the hard, rocky soil of New-Hampshire, as full of bowlders and pebbles as a Christmas pudding is of plums, plowing with such an implement was a sorry business at best. My father hitched eight oxen and a horse to his plow when he broke up pebbly green-sward, and found an acre of it a very long day's work. I hardly need add that subsoiling was out of the question, and that six inches was the average depth of his furrow.

I judge that the best Steel Plows now in use do twice the execution that his did with a like expenditure of power—that we can, with equal power, plow twelve inches as easily and rapidly as he plowed six. Ought we to do it? Will it pay?

I first farmed for myself in 1845 on a plat of eight acres, in what was then the open country skirting the East River nearly abreast the lower point of Blackwell's Island, near Fiftieth-st., on a little indentation of the shore known as Turtle Bay. None of the Avenues east of Third was then opened above Thirtieth-st.; and the neighborhood, though now perforated by streets and covered with houses, was as rural and secluded as heart could wish. One fine Spring morning, a neighbor called and offered to plow for $5 my acre of tillage not cut up by rows of box and other shrubs; and I told him to go ahead. I came home next evening, just as he was finishing the job, which I contemplated most ruefully. His plow was a pocket edition ; his team a single horse ; his furrows at most five inches deep. I paid him, but told him plainly that I would have preferred to give the money for nothing. He insisted that he had plowed for me as he plowed for others all around me. "I will tell you," I rejoined, " exactly how this will work. Throughout the Spring and early Summer, we shall have frequent rains and moderate heat: thus far, my crops will do well. But then will come hot weeks, with little or no rain; and they will dry up this shallow soil and every thing planted thereon."

The result signally justified my prediction. We had frequent rains and cloudy, mild weather, till the 1st of July, when the clouds vanished, the sun came out intensely hot, and we had scarcely a sprinkle till the 1st of September, by which time my Corn and Potatoes had about given up the ghost. Like the seed which fell on stony ground in the Parable of the Sower, that which I had planted had withered away " because there was no root ;" and my prospect for a harvest was utterly blighted, where, with twelve inches of loose, fertile, well pulverized earth at their roots, my crops would have been at least respectable. When I became once more a farmer in a small way on my present place, I had not forgotten the lesson, and I tried to have plowed deeply and thoroughly so much land as I had plowed at all. My first Summer here (1853) was a very dry one, and crops failed in consequence around me and all over the country ; yet mine were at least fair ; and I was largely indebted for them to relatively deep plowing. I have since suffered from frost (on my low land), from the rotting of seed in the ground, from the ravages of insects, etc.; but never by drouth; and I am entirely confident that Deep Plowing has done me excellent service. My only trouble has been to get it done; for there are apt to be reasons—(haste, lateness in the season, etc.)—for plowing shallowly for "just this time," with full intent to do henceforth better.

I close this paper with a statement made to me by an intelligent British farmer living at Maidstone, south of England. He said:

"A few years ago there came into my hands a field of twelve acres, which had been an orchard; but the trees were hopelessly in their dotage. They must be cut down; then their roots must be grubbed out; so I resolved to make a clean job of it, and give the field a thorough trenching. Choosing a time in Autumn or early Winter when labor was abundant and cheap, I had it turned over three spits (27 inches) deep; the lowest being merely reversed; the next reversed and placed at the top; the surface being reversed and placed below the second. The soil was strong and deep, as that of an orchard should be; I planted the field to Garden Peas, and my first picking was very abundant. About the time that peas usually begin to wither and die, the roots of mine struck the rich soil which had been the first stratum, but was now the second, and at once the stalks evinced a new life—

threw out new blossoms, which were followed by pods; and so kept on blossoming and forming peas for weeks, until this first crop far more than paid the cost of trenching and cultivation."

Thus far my English friend. Who will this year try a patch of Peas on a plat made rich and mellow for a depth of at least two feet, and frequently moistened in Summer by some rude kind of irrigation?

The fierceness of our Summer suns, when not counteracted by frequent showers, shortens deplorably the productiveness of many Vegetables and Berries. Our Strawberries bear well, but too briefly; our Peas wither up and cease to blossom after they have been two or three weeks plump enough to pick. Our Raspberries, Blackberries, etc., fruit well, but are out of bearing too soon after they begin to yield their treasures. I am confident that this need not be. With a deep, rich soil, kept moistened by a periodical flow of water, there need not and should not be any such haste to give over blooming and bearing. The fruit is Nature's attestation of the geniality of the season, the richness and abundance of the elements inhering in the soil or supplied to it by the water. Double the supply of these, and sterility should be postponed to a far later day than that in which it is now inaugurated.

XV.

PLOWING—GOOD AND BAD.

THERE are so many wrong ways to do a thing to but one right one that there is no reason in the impatience too often evinced with those who contrive to swallow the truth wrong end foremost, and thereupon insist that it won't do. For instance : A farmer hears something said of deep plowing, and, without any clear understanding of or firm faith in it, resolves to give it a trial. So he buys a great plow, makes up a strong team, and proceeds to turn up a field hitherto plowed but six inches to a depth of a foot : in other words, to bury its soil under six inches of cold, sterile clay, sand, or gravel. On this, he plants or sows grain, and is lucky indeed if he realizes half a crop. Hereupon, he reports to his neighbors that Deep Plowing is a humbug, as he suspected all along ; but now he knows, for he has tried it. There are several other wrong ways, which I will hurry over, in order to set forth that which I regard as the right one.

Here is a middling farmer of the old school, who walks carefully in the footsteps of his respected grandfather, but with inferior success, because sixty annual harvests, though not particularly luxuriant, have partially exhausted the productive capacity of the acres he inherited.   He now garners from fifteen to thirty bushels per acre of Corn, from ten to twenty of Wheat, from fifteen to twenty of Rye, from twenty to thirty of Oats, and from a tun to a tun and a half of Hay, as the season proves more or less propitious, and just contrives to draw from his sixty to one hundred acres a decent subsistence for his family; plowing, as his father and grandfather did, to a depth of five to seven inches: What can Deep Plowing do for *him?*

I answer—By itself, nothing whatever.   If in every other respect he is to persist in doing just as his father and his grandfather did, I doubt the expediency of doubling the depth of his furrows. True, the worst effects of the change would be realized at the outset, and I feel confident that his six inches of subsoil, having been made to change places with that which formerly rested upon it, must gradually be wrought upon by air, and rain, and frost, until converted into a tolerably productive soil, through which the roots of most plants would easily and speedily make their way down to the richer stratum which, originally surface, has been transposed into subsoil.   But this exchange of positions between the original surface and subsoil is not what I mean by Deep Plowing, nor anything like it. What I *do* mean is this:

Having thoroughly underdrained a field, so that

water will not stand upon any part of its surface, no matter how much may there be deposited, the next step in order is to increase the depth of the soil. To this end, procure a regular sub-soil plow of the most approved pattern, attach to it a strong team, and let it follow the breaking-plow in its furrow, lifting and pulverizing the sub-soil to a depth of not less than six inches, but leaving it in position exactly where it was. The surface-plow turns the next furrow upon this loosened sub-soil, and so on till the whole field is thus pulverized to a depth of not less than twelve inches, or, better still, fifteen. Now, please remember that you have twice as much soil per acre to fertilize as there was before; hence, that it consequently requires twice as much manure, and you will have laid a good foundation for increased crops. I do not say that all the additional outlay will be returned to you in the increase of your next crop, for I do not believe anything of the sort; but I *do* believe that this crop will be considerably larger for this generous treatment, especially if the season prove remarkably dry or uncommonly wet; and that you will have insured better crops in the years to come, including heavier grass, after that field shall once more be laid down; and that, in case of the planting of that field to fruit or other trees, they will grow faster, resist disease better, and thrive longer, than if the soil were still plowed as of old. (I shall insist hereafter on the advantage and importance of subsoiling orchards.)

Take another aspect—that of subsoiling hill-sides to prevent their abrasion by water :

I have two bits of warm, gravelly hill-side, which bountifully yield Corn, Wheat and Oats, but which are addicted to washing. I presume one of these bits, at the south-east corner of my farm, has been plowed and planted not less than one hundred times, and that at least half the fertilizers applied to it have been washed into the brook, and hence into the Hudson. To say that $1,000 have thus been squandered on that patch of ground, would be to keep far within the truth. And, along with the fertilizers, a large portion of the finer and better elements of the original soil have thus been swept into the brook, and so lavished upon the waters of our bay. But, since I had those lots thoroughly subsoiled, all the water that falls upon them when in tillage sinks into the soil, and remains there until drained away by filtration or evaporation ; and I never saw a particle of soil washed from either save once, when a thaw of one or two inches on the surface, leaving the ground solidly frozen beneath, being quickly followed by a pouring rain, washed away a few bushels of the loosened and sodden surface, proving that the law by virtue of which these fields were formerly denuded while in cultivation is still active, and that Deep Plowing is an effective and all but unfailing antidote for the evil it tends to incite.

We plow too many acres annually, and do not plow them so thoroughly as we ought. In the good time coming, when Steam shall have been so harnessed to a gang of six to twelve plows that, with one man guiding and firing, it will move as fast as a man ought to walk, steaming on and thoroughly pulverizing from twelve to twenty-five acres per day, I believe we shall plow at least two feet deep, and plow not less than twice before putting in any crop whatever. Then we may lay down a field in the confident trust that it will yield from two and a half to three tuns of good hay per annum for the next ten or twelve years; while, by the help of irrigation and occasional top-dressing, it may be made to average at least three tuns for a life-time, if not forever.

When my Grass-land requires breaking up—as it sometimes does—I understand that it was not properly laid down, or has not been been well treated since. A good grazing farmer once insisted in my hearing that grass-land should *never* be plowed—that the vegetable mold forming the surface, when the timber was first cut off, should remain on the surface forever. Considering how uneven the stumps and roots and cradle-knolls of a primitive forest are apt to leave the ground, I judge that this is an extreme statement. But land once thoroughly plowed and subsoiled ought thereafter to be kept in grass by liberal applications of Gypsum, well-cured Muck, and barn-yard Manure to its surface, without needing to be plowed again and reseeded. Put back in Manure what is taken off in Hay, and the Grass should hold its own.

XVI.

My little, hilly, rocky farm teaches lessons of thoroughness which I would gladly impart to the boys of to-day who are destined to be the farmers of the last quarter of this century. I am sure they will find profit in farming better than their grandfathers did, and especially in putting their land into the best possible condition for effective tillage. There were stones in my fields varying in size from that of a brass kettle up to that of a hay-cock—some of them raising their heads above the surface, others burrowing just below it—which had been plowed around and over perhaps a hundred times, till I went at them with team and bar, or (where necessary) with drill and blast, turned or blew them out, and hauled them away, so that they will interfere with cultivation nevermore. I insist that this is a profitable operation—that a field which will not pay for such clearing should be planted with trees and thrown out of cultivation conclusively. Dodging and skulking from rock to rock is hard upon team, plow, and plowman; and it can rarely pay. Land ribbed and spotted with fast rocks will pay if

(96)

judiciously planted with Timber—possibly if well set
in Fruit—but tilling it from year to year is a thankless task ; and its owner may better work by the day
for his neighbors than try to make his bread by such
tillage.

So with fields soaked by springs or sodden with
stagnant water. If you say you cannot afford to
drain your wet land, I respond that you can still less
afford to till it without draining. If you really cannot afford to fit it for cultivation, your next best
course is to let it severely alone.

A poor man who has a rough, rugged, sterile farm,
which he is unable to bring to its best possible condition at once, yet which he clings to and must live
from, should resolve that, if life and health be spared
him, he will reclaim one field each year until all that
is not devoted to timber shall have been brought into
high condition. When his Summer harvest is over,
and his Fall crops have received their last cultivation, there will generally be from one to two Autumn
months which he can devote mainly to this work.
Let him take hold of it with resolute purpose to improve every available hour, not by running over the
largest possible area, but by dealing with one field
so thoroughly that it will need no more during a long
life-time. If it has stone that the plow will reach,
dig them out; if it needs draining, drain it so
thoroughly that it may hereafter be plowed in
Spring so soon as the frost leaves it; and now let
soil and subsoil be so loosened and pulverized that

roots may freely penetrate them to a depth of fifteen to twenty inches, finding nourishment all the way, with incitement to go further if ever failing moisture shall render this necessary. Drouth habitually shortens our Fall crops from ten to fifty per cent.; it is sure to injure us more gravely as our forests are swept away by ax and fire; and, while much may be done to mitigate its ravages by enriching the soil so as to give your crops an early start, and a rank, luxuriant growth, the farmer's chief reliance must still be a depth of soil adequate to withstand weeks of the fiercest sunshine.

I have considered what is urged as to the choice of roots to run just beneath the surface, and it does not signify. Roots seek at once heat and moisture; if the moisture awaits them close to the surface, of course they mainly run there, because the heat is there greatest. If moisture fails there, they must descend to seek it, even at the cost of finding the heat inadequate—though heat increases and descends under the fervid suns which rob the surface of moisture. Make the soil rich and mellow ever so far down, and you need not fear that the roots will descend an inch lower than they should. *They* understand their business; it is *your* sagacity that may possibly prove deficient.

I suspect that the average farmer does far too little plowing—by which I mean, not that he plows too few acres, for he often plows too many, but that he should plow oftener as well as deeper and more thoroughly. I spent three or four of my boyish Summers planting and tilling Corn and Potatoes on fields broken up just before they were planted, never cross-plowed, and of course tough and intractable throughout the season. The yield of Corn was middling, considering the season ; that of Potatoes more than middling ; yet, if those fields had been well plowed in the previous Autumn, cross-plowed early in the Spring, and thoroughly harrowed just before planting-time, I am confident that the yield would have been far greater, and the labor (save in harvesting) rather less—the cost of the Fall plowing being over-balanced by the saving of half the time necessarily given to the planting and hoeing.

Fall Plowing has this recommendation—it lightens labor at the busier season, by transfering it to one of comparative dullness. I may have said that I consider him a good farmer who knows how to make a rainy day equally effective with one that is dry and fair ; and, in the same spirit, I count him my master in this art who can make a day's work in Autumn or Winter save a day's work in Spring or Summer. Show me a farmer who has no land plowed when May opens, and is just waking up to a consciousness that his fences need mending and his trees want trimming, and I will guess that the sheriff will be after him before May comes round again.

There is no superstition in the belief that land is (or may be) enriched by Fall Plowing. The Autumn gales are freighted with the more volatile elements of decaying vegetation. These, taken up wherever they are given off in excess, are wafted to and deposited in the soils best fitted for their reception. Regarded simply as a method of fertilizing, I do not say that Fall Plowing is the cheapest; I *do* say that any poor field, if well plowed in the Fall, will be in better heart the next Spring, for what wind and rain will meantime have deposited thereon. Frost, too, in any region where the ground freezes, and especially where it freezes and thaws repeatedly, plays an important and beneficial part in aerating and pulverizing a freshly plowed soil, especially one thrown up into ridges, so as to be most thoroughly exposed to the action of the more volatile elements. The farmer who has a good team may profitably keep the plow running in Autumn until every rood that he means to till next season has been thoroughly pulverized.

In this section, our minute chequer-work of fences operates to obstruct and impede Plowing. Our predecessors wished to clear their fields, at least superficially, of the loose, troublesome bowlders of granite wherewith they were so thickly sown; they mistakenly fancied that they could lighten their own toil by sending their cattle to graze, browse, and gnaw, wherever a crop was not actually on the ground; so they fenced their farms into patches of two or ten acres, and thought they had thereby increased their value! That was a sad miscalculation. Weeds, briars and bushes were sheltered and nourished by

these walls; weasels, rats and other destructive animals, found protection and impunity therein ; a
wide belt on either side was made useless or worse;
while Plowing was rendered laborious, difficult, and
inefficient, by the necessity of turning after every
few hundred steps. We are growing slowly wiser,
and burying a part of these walls, or building them
into concrete barns or other useful structures; but
they are still far too plentiful, and need to be dealt
with more sternly. O squatter on a wide prairie, on
the bleak Plains, or in a broad Pacific valley, where
wood must be hauled for miles and loose stone are
rarely visible, thank God for the benignant dispensation which has precluded you from half spoiling your
farm by a multiplicity of obstructing, deforming
fences, and so left its soil free and open to be every
where pervaded, loosened, permeated, by the reno-
vating Plow!

XVII.

PRICES vary so widely in different localities that no fertilizer can be pronounced everywhere cheapest or best worth buying; and yet I doubt that there is a rood of our country's surface in fit condition to be cultivated to which Gypsum (Plaster of Paris) might not be applied with profit. Where it costs $10 or over per tun, I would apply it sparingly—say, one bushel per acre—while I judge three bushels per acre none too much in regions where it may be bought much cheaper. Even the poor man who has but one cow, should buy a barrel of it, and dust his stable therewith after cleaning it each day. He who has a stock of cattle should never be without it, and should freely use it, alike in stable and yard, to keep down the noisome odors, and thus retain the volatile elements of the manure. Every meadow, every pasture, should be sown with it at least triennially; where it is abundant and cheap, as in Central New-York, I would apply it each year, unless careful observation should satisfy me that it no longer subserved a good purpose.

As to the *time* of application, while I judge any

season will do, my present impression is that it will do most good if applied when the Summer is hottest and the ground driest. If, for instance, you close your haying in mid-Summer, having been hurried by the rapid ripening of the grass, and find your meadows baked and cracked by the intense heat, I reckon that you may proceed to dust those meadows with Gypsum with a moral certainty that none of it will be wasted. So if your Corn and other Fall crops are suffering from and likely to be stunted by drouth, I advise the application of Gypsum broadcast, as evenly as may be and as bounteously as its price and your means will allow. I do not believe it so well to apply it specially to the growing stalks, a spoon-full or so per hill; and I doubt that it is ever judicious to plant it *in* the hill with the seed. The readiest and quickest mode of application is also, I believe, the best.

How Gypsum impels and invigorates vegetable growth, I do not pretend to know; but that it *does* so was demonstrated by Nature long before Man took the hint that she freely gave. The city of Paris and a considerable adjacent district rest on a bed of Gypsum, ranging from five to twenty feet below the surface, and considerably decomposed in its upper portion by the action of water. This region produces Wheat most luxuriantly, and I presume has done so from time immemorial. At length it crawled through the hair of the tillers of this soil that the substance which did so much good fortuitously, and (as it were) because it could not do otherwise, might do still more if applied to the soil, with deliberate intent to test its value as a fertilizer. The result we all understand.

Gypsum is a chemical compound of Sulphur and Lime—so much is agreed; and the theory of chemists has been that, as the winds pass over a surface sown with it, the Ammonia which has been exhaled by a thousand barn-yards, bogs, &c., having a stronger affinity for Sulphur than Lime has, dissolves the Gypsum, combines with the Sulphur, forming a Sulphate of Ammonia, and leaves the Lime to get on as it may. I accept this theory, having no reason to distrust it; and, knowing that Sulphate of Ammonia is a powerful stimulant of vegetable growth (as any one may be assured by buying a little of it from some druggist and making the necessary application), I can readily see how the desired result *might* in this way be produced. For our purpose, however, let it suffice that *it is* produced, of which almost any one may be convinced by sowing with Gypsum and passing by alternate strips or belts of the·same clover-field. I suspect that not many fertilizers repay their cost out of the first crop; but I account Gypsum one of them; and I submit that no farmer can afford not to try it. That its good effect is diminished by many and frequent applications, is highly probable; but there is no hill or slope to which Gypsum has never yet been applied which ought not to make its acquaintance this very year. I am confident that there are pastures which might be made to increase their yield of Grass one-third by a moderate dressing of it.

I have heard Andrew B. Dickinson, late of Steuben County, and one of the best unscientific, unlearned farmers ever produced by our State, maintain that he can not only enrich his own farm but impoverish his neighbors' by the free use of Gypsum on his woodless hills. The chemist's explanation of this effect is above indicated. The plastered land attracts and absorbs not only its own fair proportion of the breeze-borne Ammonia, but much that, if the equilibrium had not been disturbed by such application, would have been deposited on the adjacent hills. As Mr. D. makes not the smallest pretensions to science, the coincidence between his dictum and the chemist's theory is noteworthy.

Now that our country is completely gridironed with Canals and Railroads, bringing whatever has a mercantile value very near every one's door, I suggest that no township should go without Gypsum. Five dollars will buy at least two barrels of it almost anywhere; and two barrels may be sown over five or six acres. Let it be sown so that its effect (or non-effect) may be palpable; give it a fair, careful trial, and await the result. If it seem to subserve no good purpose, be not too swift to enter up judgment; but buy two barrels more, vary your time and method of application, and try again. If the result be still null, let it be given up that Gypsum is not the fertilizer needed just there—that some ill-under-

*

stood peculiarity of soil or climate renders it there ineffective. Then let its use be there abandoned; but it will still remain true that, in many localities and in countless instances, Gypsum has been fully proved one of the best and cheapest commercial fertilizers known to mankind.

I never tried, but on the strength of others' testimony believe in the improvement of soils by means of calcined clay or earth. Mr. Andrew B. Dickinson showed me where he had, during a dry Autumn plowed up the road-sides through his farm, started fires with a few roots or sticks, and then piled on sods of the upturned clay and grass-roots till the fire was nearly smothered, when each heap smoked and smouldered like a little coal-pit till all of it that was combustible was reduced to ashes, when ashes and burned clay were shoveled into a cart and strewn over his fields, to the decided improvement of their crops. Whoever has a clay sod to plow up, and is deficient in manure, may repeat this experiment with a moral certainty of liberal returns.

XVIII.

I DO not know a rood of our country's surface so rich in *all* the materials which enter into the production of the Grains, Grasses, Fruits, and Vegetables, which are the objects and rewards of cultivation, that it could not be improved by the application of fertilizers; if there be such, I heartily congratulate the owners, and advise them not to sell. Nor do I believe that there are many acres so fertile that they would not produce more Indian Corn, more Hemp, more Cotton, and more of whatever may be their appropriate staple, if judiciously fertilized. If there be farms or fields originally so good that manure would not increase their yield, I am confident that the first half-dozen crops will have taken that conceit out of them. Prairies and river-bottoms may yield ever so bounteously; but that very luxuriance of growth insures their gradual exhaustion of certain elements of crops, which must needs be replaced or their product will dwindle. Whoever has sold a thousand bushels of grain, or its equivalent in meat, from his farm, has thereby impoverished that farm, unless he has ap-

plied something that balances its loss. "I perceive
that virtue has gone out of me," observed the Saviour,
because the hem of his garment had been touched;
and every field that had been cropped might make a
similar report whenever its annual loss by abstrae-
tion has not been balanced by some kind of fertilizer.
The farmer who grows the largest crops is the most
merciless exhauster of the soil, unless he balances his
annual drafts (as good farmers rarely fail to do) by at
least equal reënforcements of the productive capacity
of his fields.

The good farmer begins by inquiring, "Wherein
was my soil originally deficient? and of what has it
been exhausted by subsequent crops?" I judge that
my gravelly hill-sides would reward the application
of two hundred loads (or tuns) of pure clay per acre,
as I think the clay flats which border Lake Champlain would pay for a like application of sand or fine
gravel where that material is found in convenient
proximity; and yet I know very well that, on at
least three-fourths of our country's area, such application would cost far more than it would be worth.
Every farmer must act on his knowledge of his soil
and its peculiar needs, and not blindly follow the dictum of another. Yet I know few farms which, were
they mine, I would not consider enhanced in value
by a vigorous application of *some* alkaline substance
—Lime, Salt, Ashes, or some of the cheaper Nitrates.
I should be very glad to apply one thousand bushels
of good house-made, hard-wood Ashes to my twenty

acres of arable upland, if I could buy them, delivered, at twenty-five cents per bushel; but they are not to be had. I doubt that there are a hundred acres of warm, dry, gravelly or sandy soil east of the Alleghanies that would not amply reward a similar application. But Ashes in quantity are unattainable, since no good farmer sells them, and Coal is the chief fuel of cities and villages. The Marls of New-Jersey I judge fully equal in average value to Ashes which have been nearly deprived of their potash by leaching, but not quite half equal, bushel for bushel, to *un*leached Ashes. I judge that average Marl is worth 10 cents per bushel where Ashes may be had for 25. But Marl is found only in a few localities, and a material worth but 10 cents per bushel will not bear transportation beyond 40 miles by wagon or 200 by water. Salt is only found or made at a few points, and is too dear for general use as a fertilizer. Where the refuse product of Salt-Works can be cheaply bought, good farmers will eagerly compete for it, if their lands at all resemble mine. I judge the tun of Potash I ordered fifteen years ago from Syracuse, paying $50 and transportation, was the cheapest fertilizer I ever bought. It was so impregnated with Salt (from the boiling over of the salt-kettles into the ashes) as to be worthless for other than agricultural purposes; but I mixed it with a large pile of Muck that I had recently dug, and, six or eight months thereafter, applied the product to a very poor, gravelly hill-side which I had just broken

up; and the immediate result was a noble crop of
Corn.   That hill-side has not yet forgotten the application.

—If I should try to explain just how and why Lime
is a fertilizer, I should probably fail; and I am well
assured that liming has in some cases been overdone;
yet I think most observers will concur in my statement that *any region which has been limed year after
year produces crops of noticeable excellence.*   I cite as
examples Chester and Lancaster Counties, Pennsylvania, with Stark and adjacent counties of Ohio.
Possibly, results equally gratifying might be secured
by applying some other substance; I only *know* that
frequently limed lands are generally good lands, as
their crops do testify..   I heartily wish that the flat
clay intervales of Western Vermont could have a
fair trial of the virtues of liming.   I should expect
to see them thereby rendered friable and arable; no
longer changing speedily from the semblance of tar
to that of brick, but readily plowed and tilled, and
yielding liberally of Grain as well as Grass.   I am
confident that most farms in our country will pay
for liming to the extent of fifty bushels per acre
where the cost of quick-lime does not exceed ten
cents per bushel; and most farmers, by taking, hot
from the kiln, the refuse lime that is deemed unfit
for building purposes, can obtain it cheaper than that.

I wish some farmer who gives constant personal
attention to his work—as I cannot—would make
some careful tests of the practical value of alkalis,

For instance: the abundance and tenacity of our common sorrel is supposed to indicate an acid condition of the soil ; and all who have tried it know that sorrel is hard to kill by cultivation. I suggest that whoever is troubled with it should cover two square rods with one bushel of quick-lime just after plowing and harrowing this Spring ; then apply another bushel to *four* square rods adjacent; then make similar applications of ashes to two and four square rods respectively, taking careful note of the boundaries of each patch, and leaving the rest of the field destitute of either application. I will not anticipate the result : more than one year may be required to evolve it ; but I am confident that a few such experiments would supply data whereof I am in need ; and there are doubtless others whose ignorance is nearly equal to mine.

Many have applied Lime to their fields without realizing any advantage therefrom. In some cases, there was already a sufficiency of this ingredient in the soil, and the application of more was one of those many wasteful blunders induced by our ignorance of Chemistry. But much Lime is naturally adulterated with other minerals, especially with Manganese, so that its application to most if not to all soils subserves no good end. In the absence of exact, scientific knowledge, I would buy fifty bushels of quick-lime, apply them to one acre running through a field, and watch the effect. If it does n't pay, you have a bad article, or your soil is not deficient in Lime.

XIX.

A FARMER is a manufacturer of articles wherefrom mankind are fed and clad; his raw materials are the soil and the various substances he mingles therewith or adds thereto in order to increase its productive capacity. His art consists in transforming by cultivation crude, comparatively worthless, and even noxious, offensive materials into substances grateful to the senses, nourishing to the body, and sometimes invigorating, even strengthening, to the mind.

I have heard of lands that were naturally rich enough; I never was so lucky or perchance so discerning as to find them. Yet I have seen Illinois bottoms whereof I was assured that the soil was fully sixteen feet deep, and a rich, black alluvium from top to bottom; and I do not question the statements made to me from personal observation that portions of the strongly alkaline plain or swale on which Salt Lake City is built, being for the first time plowed, irrigated, and sown to Wheat, yielded ninety bushels of good grain per acre. I never saw, yet on evidence believe, that pioneer settlers of the Miami Valley, wishing, some years after settling there, to sell their farms, advertised them as peculiarly desirable in that the barns stood over a creek or "branch," which swept away the manure each Winter or Spring without trouble to the owner; and I have myself grown both Wheat and Oats that were very rank and heavy in straw, yet which fell so flat and lay so dead that the heads scarcely bore a kernel. Had I been a wiser, better farmer, I should have known how to stiffen the straw and make it do its office, in spite of wind and storm.

[And let me here say, lest I forget it in its appropriate place, that I am confident that most farmers sow grain too thickly for any but very poor land. If one thinks it necessary to scatter three bushels of Oats per acre, I tell him that he should apply more manure and less seed—that land which requires three bushels of seed is not rich enough to bear Oats. He might better concentrate his manure on half so much land, and save two-thirds of his seed.]

I do not hold that the remarkably rich soils I have instanced needed fertilizing when first plowed; I will presume that they did not. Yet, having never yet succeeded in manuring a corn-field so high that a few loads more would not (I judge) have increased the crop, I doubt whether even the richest Illinois bottoms would not yield more Corn, year by year, if reënforced with the contents of a good barn-yard. And, when the first heavy crop of Corn has been taken from a field, that field—no matter how deep and fertile its soil—is less rich in corn-forming elements than it was before. Just so sure as that there is no depletion or shrinkage when nothing is taken from nothing, so sure is it that something cannot be taken from something without diminishing its capacity to yield something at the next call. Rotation of crops is an excellent plan ; for one may flourish on that which another has rejected ; but this does not over-bear Nature's inflexible exaction of so much for so much. Hence, if there ever was a field so rich that nothing could be added that would increase its productive capacity, the first exacting crop thereafter taken from it diminished that capacity, and rendered a fresh application of some fertilizer desirable.

Years ago, a Western man exhibited at our Farmers' Club a specimen of the soil of his region which was justly deemed very rich, taken from a field whereon Corn had been repeatedly grown without apparent exhaustion. A chemical analysis had been made of it, which was submitted with the soil. It was claimed that nothing could improve its capacity for producing the great Illinois staple. Prof, Mapes dissented from this conclusion.· "This soil," said he, "while very rich in nearly every element which enters into the composition of Corn, gives barely a trace of Chlorine, the base of Salt. Hence, if five bushels per acre of Salt be applied to that field, and it does not thereupon yield five bushels more per annum of Corn, I will agree to eat the field."

Many men fertilize their poor lands only, supposing that the better can do without. I judge that to be a mistake. My rule would be to plant the poorest with such choice trees as thrive without manure, and pile the fertilizers upon the better. It seems to me plain that of two fields, one of which has a soil containing nine-tenths of the elements of the desired crop, while the other shows but one to three-tenths, it is a more hopeful and less thankless task to enrich the former than the latter. If you are required to supply to a field nearly everything that your proposed crop will withdraw from it, I do not see where the profit comes in ; but if you are required to supply but a tenth, because the soil as you found it stood ready to contribute the remaining nine-tenths, it seems to me that the margin for profit is here decidedly the greater.

How many tuns of earth ought a farmer to be obliged to turn over and over in order to obtain therefrom a hundred bushels of Corn? Two hundred? Five hundred? A thousand? *Five* thousand? Other things being equal, no one will doubt that, if he can make the Corn from one hundred tuns of soil, it were better to do so than to employ five hundred or five thousand. It seems clear to my mind that, though other conditions be *un*equal, it is generally well to endeavor to produce the required quantity from the smaller rather than the larger area.

I fully share the average farmer's partiality for barn-yard manure in preference to most, if not all, commercial fertilizers. In my judgment, almost any farmer who has cattle, with fit shelter and Winter fodder, can make fertilizers far cheaper than he can buy them. I judge that almost every farmer who has paid $100 or over for Guano (for instance), might have more considerably enriched his farm by draw-ing muck from some convenient bog or pond into his barn-yard in August or September and carting it thence to his fields the next Fall. If he can get no muck within a mile, let him cut, when they are in blossom, all the weeds that grow near him, es-pecially by the road-side, cart them at once into his barn-yard, and there convert them into fertilizers. In Autumn, replace the hay-rack on the wagon or cart, and pile load after load of freshly-fallen leaves into your yard; taking them, if you may, from the sides of roads and fences, and from any place where they may have been lodged or heaped by the winds, your own wood-lot excepted. Plow the turf off of any scurvy lot or road-side, and pile it into the barn-yard; nay, dig a hundred loads of pure clay, and place it there, if you can get it at a small expense, and your average soil is gravelly or sandy. The farmer who is unable or reluctant to buy commercial fertil-izers should apply his whole force every Autumn to replenishing his barn-yard with that material which he can obtain most easily which the trampling of his cattle may readily convert into manure. A month is too little, two months would not be too much, to de-vote to this good work. Some may seem obliged to postpone it to Winter; but that is to run the risk of embarrassment by frost or snow, and encounter the certainty that your material will be inferior in quality, or not so well fitted to apply to grain-crops the ensuing Fall.

—All this, you may say, is not instruction. We ought to know exactly what lands are enriched by Gypsum, and what, if any, are not; why these are fertilized, why those are not, by a common application; how great is the profit of such application in any case; and what substitute can most nearly subserve the same ends where Gypsum is not to be had. I admit all you claim, and do not doubt that there shall yet be a Scientific Agriculture that will fully answer your requirements. As yet, however, it exists but in suggestions and fragments; and attempts to complete it by naked assertions and sweeping generalizations tend rather to mislead and disgust the young farmer than really to enlighten and guide him. At all events, I shall aim to set forth as true no more than I know, or with good reason confidently believe.

I close by reiterating my belief that no farmer ever yet impoverished himself by making too much manure or by applying too much of his own manufacture. I cannot speak so confidently of *buying* commercial fertilizers; but these I will discuss in my next chapter.

XX.

I HATE to check improvement or chill the glow of Faith; yet I do so keenly apprehend that many of our people, especially among the Southern cotton-growers, are squandering money on Commercial Fertilizers, that I am bound to utter my note of warning, even though it should pass wholly unheeded. Let me make my position as clear as I can.

I live in a section which has been cultivated for more than two centuries, while its proximity to a great city has tempted to crop it incessantly, exhaustively. Wheat while its original surface soil of six to twelve inches of vegetable mold (mainly composed of decayed forest-leaves) remained; then Corn and Oats; at length, Milk, Beef, and Apples—have exhausted the hill-sides and gentler slopes of Westchester County, except where they have been kept in heart by judicious culture and liberal fertilizing; and, even here, that subtle element, Phosphorus, which enters minutely but necessarily into the composition of every animal and nearly every vegetable structure, has been gradually drawn away in Grain,

(118)

in Milk, in Bones, and not restored to the soil by the application of ordinary manures. I am convinced that a field may be so manured as to give three tuns of Hay per acre, yet so destitute of Phosphorus that a sound, healthy animal cannot be grown therefrom. For two centuries, the tillers of Westchester County knew nothing of Chemistry or Phosphorus, and allowed the unvalued bones of their animals to be exported to fatten British meadows, without an effort to retain them. Hence, it has become absolutely essential that we buy and apply Phosphates, even though the price be high; for our land can no longer do without them. Wherever a steer or heifer can occasionally be caught gnawing or mumbling over an old bone, there Phosphates are indispensable, no matter at what cost. Better pay $100 per tun for a dressing of one hundred pounds of Bone per acre than try to do without.

But no lands recently brought into cultivation—no lands where the bones of the animals fed thereon have been allowed, for unnumbered years past, to mingle with the soil—can be equally hungry for Phosphates; and I doubt that any cotton-field in the South will ever return an outlay of even $50 per tun for any Phosphatic fertilizer whatever. That *any* preparation of Bone, or whereof Bone is a principal element, will increase the succeeding crops, is undoubted; but that it will ever return its cost and a decent margin of profit, is yet to be demonstrated to my satisfaction.

No doubt, there are special cases in which the application even of Peruvian Guano at $90 per tun is advisable. A compost of Muck, Lime, &c., equally efficient, might be far cheaper; but months would be required to prepare and perfect it, and meantime the farmer would lose his crop, or fail to make one. If a tun of Guano, or of some expensive Phosphate, will give him six or eight acres of Clover where he would otherwise have little or none, and he needs that Clover to feed the team wherewith he is breaking up and fitting his farm to grow a good crop next year, he may wisely make the purchase and application, even though he may be able to compost for next year's use twice the value of fertilizers for the precise cost of this. But I am so thorough in my devotion to "home industry," that I hold him an unskillful farmer who cannot, nine times in ten, make, mainly from materials to be found on or near his farm, a pile of compost for $100 that will add more to the enduring fertility of his farm than anything he can bring from a distance at a cost of $150.

Understand that this is a general rule, and subject, like all general rules, to exceptions. Gypsum, I think every farmer should buy; Lime also, if his soil needs it; Phosphates in some shape, if past ignorance or folly has allowed that soil to be despoiled of them; Wood Ashes, if any one can be found so brainless as to sell them; Marl, of course, where it is found within ten miles; Guano very rarely, and mainly when something is needed to make a crop before coarser and colder fertilizers can be brought into a condition of fitness for use; but the general rule I insist on is this: A good farmer will, in the course of twenty or thirty years, make at least $10 worth of fertilizers for every dollar's worth he buys from any dealer, unless it be the sweepings or other excretions of some not distant city.

I have used Guano frequently, and, though it has generally made its mark, I never yet felt sure that it returned me a profit over its cost. Phosphates have done better, especially where applied to Corn in the hill, either at the time of planting or later; yet my strong impression is that Flour of Bone, applied broadcast and freely, especially when Wheat or Oats are sown on a field that is to be laid down to Grass, pays better and more surely than anything else I order from the City, Gypsum, and possibly Oyster-Shell Lime, excepted.

My experience can be no safe guide for others, since it is not proved that the anterior condition and needs of their soils are precisely like those of mine. I apprehend that Guano has not had a fair trial on my place—that carelessness in pulverizing or in application has caused it to " waste its sweetness on the desert air," or that a drouth following its application has prevented the due development of its virtues. And still my impression that Guano is the brandy of vegetation, supplying to plants stimulus rather than nutrition, is so clear and strong that it may not easily be effaced. It seems to me plainly absurd to send ten thousand miles for this stimulant, when this or any other great city annually poisons its own atmosphere and the adjacent waters with excretions which are of very similar character and value, and which Science and Capital might combine to utilize at less than half the cost of like elements in the form of Guano.

My object in this paper is to incite experiment and careful observation. No farmer should absolutely trust aught but his own senses. A Rhode Islander once assured me that he applied to four acres of thin, slaty gravel one hundred pounds per acre of Nitrate of Soda which cost him $4 per hundred, and obtained therefrom four additional tuns of good Hay, worth $15 per tun : Net profit (after allowing for the cost of making the Hay), say $30. He might not be so fortunate on a second trial, and there may not be another four acres of the earth's surface where Nitrate of Soda would do so well; but, should I ever have a fair opportunity, I mean to see what a little of that Nitrate will do for *me.* And I hope farmers may more and more be induced to conform in practice to the Apostolic precept, "Prove all things: Hold fast that which is good." No one's success or failure in a particular instance should be conclusive with others, because of the infinite diversity of antecedent and attendant circumstances; but if every thrifty farmer would give to each of the commercial fertilizers—Lime, Gypsum, Guano, Raw Bone, Phosphates, Ashes, Salt, Marl, etc.—such a careful trial as he might, observing closely and recording carefully the results, we should soon have a mass of facts and results, wherefrom deductions might be drawn of signal practical value to the present and to future generations.

I firmly believe that great results of signal beneficence are to be slowly but surely achieved by means of the household convenience known as the Earth-Closet, and by kindred devices for rendering inoffensive and utilizing the most powerful fertilizer produced on every farm and in every household. That is a vulgar squeamishness which leaves it to poison the atmosphere and offend the senses on the assumption that it is too noisome to be dealt with or utilized. A true refinement counsels that it be daily covered, and its odor absorbed or suppressed by earth, or muck, or ashes, and thus prepared for removal to and incorporation with the soil. It is far within the truth to estimate our National loss by the waste of this material at $1 per head, or $40,000,000 in all per annum : a waste which is steadily diminishing the productive capacity of our soil. This cannot, must not, be allowed to continue. We must devise or adopt *some* mode of securing and applying this powerful fertilizer ; and I defer to that which is already in extensive and daily expanding use. Let whoever can do better ; but meantime let us welcome and diffuse the Earth Closet.

XXI.

MUCK—HOW TO UTILIZE IT.

THE time will be, I cannot doubt, when chemists can tell us the exact positive or relative value of a cord of Muck—how this swamp or that pond affords a choice article, while the product of another will hardly pay for digging. There may be chemists whose judgment on these points is now worth far more than mine, since mine is worth exactly nothing. I *do* know, however, that Muck is a valuable fertilizer, and that digging and composting it *does pay*. I judge that I have transferred at least three thousand loads of it from my swamp to my upland; and the effect has been all that I expected. Let me speak of Muck generally, in the light of my own experience.

Wherever rocks in ridges come to the surface of a valley, plain, or gentle slope, water is apt to be collected or retained by them, forming ponds or shal lower pools, which may or may not dry up in Sum mer, but which are seldom dry late in Autumn, when plants are dying and leaves are falling. The latter, caught in their descent by the harsh winds of the season, are swept along the bare, dry ground, till they strike the water, which arrests their progress and soon engulfs them.   Thus an acre of watery surface will often collect and retain the dead foliage of five to ten acres of forest; and next Fall will render its kindred tribute, and the next, and the next, for ever.   There cannot be less than fifty millions of acres of Swamps in our old States (including Maine); whereof I presume the larger area was covered with water until the slow contributions of leaves and weeds filled them above the level at which water is no longer retained on the surface.   And still, they are so moist and boggy, and their rank vegetation is so retentive, that the leaves swept in from the adjacent hills and glades are firmly retained and aid to increase the depth of their vegetable mold, which varies from a few inches to twenty and even thirty feet.   In my old County of Westchester, I roughly estimate that there are at least five thousand acres of bog, whereof but a very few hundreds have yet been subdued to the uses of cultivation.

Whoever digs a quantity of Swamp Muck and applies it *directly* to his fields or garden, will derive little or no immediate benefit therefrom.   It is green, sour, cold, and more likely to cover his farm thickly and persistently with Sorrel, Eye-smart, Rag-weed, Pursley, and other infestations, than to add a bushel per acre to his crop of Grain or Roots.   And thus many have tried Muck, and, on trial, pronounced it a pestilent humbug.

But let any farmer turn his whole force into a bog
or marsh directly after finishing his Summer harvest
(when it is apt to be driest and warmest), and, having freed it of water to the best of his ability, dig and
draw out one hundred cords of its black, oozy substance, and he will know better than to unite in that
hasty judgment. If the bog be near his farm-yard,
let the Muck be shoveled at once into a cart and
drawn thither; but, if not, let it be simply brought
out in wheel-barrows and deposited, not more than
two feet deep, on the most convenient bank that is
well drained and perfectly dry. Here let it dry and
drain till after Fall harvest, and then begin to draw
it gradually into the yards, and especially where it
may be worked over by swine and scratched over for
seeds and insects by fowls. Assuming that the farm-
yard is lowest in the centre and allows no liquid to
escape save by evaporation, the Muck may well be
dumped on the drier sides; thence, after being
worked over and trampled through and through, to
be shoveled into the centre and replaced by fresh
arrivals. A hundred cords may thus be so mixed and
ripened as to be fit to draw out next May and used
as a fertilizer for Grain or Roots, though, if not so
treated, it should lie exposed to sun and wind a full
year; being applied in the Fall to crops of Winter
grain or spread upon the fields to be planted or sowed next Spring. All the manure made during the
Winter should be spread over that which lies in the
yard at least monthly; and then new Muck drawn in,

to be rooted or scratched over, trampled into the underlying strata, and overspread in its turn. Thus treated, I am confident that each hundred cords of Muck will be equal in value to an equal quantity of manure, though it may not give up its fertilizing properties so freely to the first crop that follows its application. I have land that did not yield (in pasture) the equivalent of half a tun of hay per annum when I bought it, that now yields at least three tuns of good hay per annum ; and its renovation is mainly due to a free application of Swamp Muck.

To those who have a good stock of animals, with Muck convenient to their yards, I would not recommend any other treatment than the foregoing; but there are many who keep few animals, or whose muck-beds lie at the back of their farms, two or three hundred rods from their barns; while they wish to fertilize the fields in this quarter, which have been slighted in former applications, because of the distance over which manure had to be hauled. If these possess or can buy good hard-wood, house-made Ashes at twenty-five cents or less per bushel, I would say, Mix these well, at the rate of two or three bushels to the cord, with your Muck as you dig it ; work it over the next Spring, and apply it the ensuing Fall, so as to give it a full year to ripen and sweeten, and it will be all right. But, if you have not and cannot get the Ashes, and *can* procure dirty, refuse Salt from some meat-packer or wholesale grocer, apply this as you would have applied the Ashes, but in rather larger quantity; and, if you can get neither Ashes nor Salt, use quick Lime, as fresh and hot from the kiln as you can apply it. The best Lime is that from burned Oyster-Shells; I consider this, if nowise slaked, nearly equal to refuse Salt; but Oyster-Shell Lime is too dear at most inland points; and here the refuse of the kilns—that which is not good enough for mason-work—must be used. Usually, the lime-burner has a load or more of this at the clearing out of every kiln, which he will sell quite cheap if it be taken out of his way at once; and this should be looked for and secured. Being inferior in quality (often because imperfectly burned), it should be applied in larger quantity—not less than four bushels to each cord of Muck.

I will not here describe the process of mixing Salt with Lime commended by Prof. Mapes, because it is not easy to bring these two ingredients together so as to mix them with the Muck as it is dug: and, though I have used them after Prof. Mapes's recipe, and purpose to do so hereafter, I do not feel certain that any positive advantage results from their blended application as a Chloride of Lime. If I should gain further light on this point before completing this series, I shall not fail to impart it.

XXII.

IF I were to estimate the average absolute loss of the farmers of this country from Insects at $100,000,000 per annum, I should doubtless be far below the mark. The loss of fruit alone by the devastations of insects, within a radius of fifty miles from this City, must amount in value to Millions. In my neighborhood, the Peach once flourished, but flourishes no more, and Cherries have been all but annihilated. Apples were till lately our most profitable and perhaps our most important product; but the worms take half our average crop and sadly damage what they do not utterly destroy. Plums we have ceased to grow or expect; our Pears are generally stung and often blighted; even the Currant has at last its fruit-destroying worm. We must fight our paltry adversaries more efficiently, or allow them to drive us wholly from the field.

Now, I have no doubt that our best allies in this inglorious warfare are the Birds. They would save us, if we did not destroy them. The British plowman, turning his sod with a myriad of crows, blackbirds, etc., chasing his steps and all but getting under his feet in their eager quest of grubs, bugs, etc., is a spectacle to be devoutly thankful for. Whenever clouds of birds shall habitually darken our fields in May and (less notably) throughout the Summer months, we may reasonably hope to grow fair crops of our favorite Fruits from year to year, and realize that we owe them to the constant, and zealous, though not quite disinterested, efforts of our friends, the Birds.

But I do not regard the ravages of Insects as entirely due to the reckless destruction and consequent scarcity of our Birds. I hold that their multiplication and their devastations are largely incited by the degeneracy of our plants caused by the badness of our culture. On this point, consider a statement made to me, some fifteen or twenty years ago, by the late Gov. William F. Packer, of Pennsylvania:

"I know (said Gov. P.) the narrow valley of a stream that runs into the west branch of the Susquehanna, which was cleared of the primitive forest some forty or fifty years since, and has ever since been alternately in tillage and grass. A road ran through the middle of it, dividing it into two narrow fields. A few years ago, this road was abandoned, and the whole of this little valley, including the road-way, thrown into a single field, which was thereupon sown to Wheat. At harvest-time, this remarkable phenomenon was presented: A good crop of sound grain on the strip four or five rods wide formerly covered by the road; while nearly every berry on either side of it was destroyed by the weevil or midge."

Now I do not infer from this fact that insect ravages are *wholly* due to our abuse and exhaustion of the soil. I presume that Wheat and other crops would be devastated by insects if there were no slovenly, niggard, exhausting tillage. But I do firmly hold that at least half our losses by insects would be precluded if our fields were habitually kept in better heart by deep culture, liberal fertilizing, and a judicious rotation of crops. I heard little of insect ravages in the wheat-fields of Western New-York throughout the first thirty years of this century; but, when crop after crop of Wheat had been taken from the same fields until they had been well nigh exhausted of their Wheat-forming elements, we began to hear of the desolation wrought by insects; and those ravages increased in magnitude until Wheat-culture had to be abandoned for years. I believe that we should have heard little of insects had Wheat been grown on those fields but one year in three since their redemption from the primal forest.

But, whatever might once have been, the Philistines are upon us. We are doomed, for at least a generation, to wage a relentless war against insects multiplied beyond reason by the neglect and shortcomings of our predecessors. We are in like condition with the inhabitants of the British isles a

thousand years ago, whose forefathers had so long
endured and so unskillfully resisted invasion and
spoliation by the Northmen that they had come to
be regarded as the sea-kings' natural prey. For
generations, it has been customary hereabout to
slaughter without remorse the birds, and let caterpillars, worms, grasshoppers, etc., multiply and ravage
unresisted. We must pay for past errors by present
loss and years of extra effort. And, precisely because
the task is so arduous, we ought to lose no time in ad-
dressing ourselves to its execution.

The first step to be taken is very simple. Let every
farmer who realizes the importance and beneficence
of Birds teach his own children and hirelings that,
except the Hawk, they are to be spared, protected,
kindly treated, and (when necessary) fed. They are
to be valued and cherished as the voluntary police of
our fields and gardens, constantly employed in fighting our battles against our ruthless foes. The boy
who robs a bird's nest is robbing the farmer of a
part of his crops. He who traverses a farm shooting
and mangling its feathered sentinels diminishes its
future product of Grain and nearly destroys that of
Fruit. The farmer might as well consent that any
strolling ruffian should shoot his Horses or Cattle as
his Birds. Begin at home to make this truth felt
and respected, and it will be the easier to impress it
also on your neighbors.

Next, there should be neighborhood or township
associations for the protection of insect-eating birls.

We must not merely agree to let them live—we must cherish and protect them. I believe that very simple cups or bowls of cast-iron, having each a hole in its centre of suitable size, that need not cost sixpence each, and could be fastened to the side of a tree with one nail lightly driven, would in time be adopted by many birds as nesting strongholds, whence they might laugh to scorn their predacious enemies. If every harmless bird could build its nest among us in a place where its eggs would be safe from hawks, crows, cats, boys, and other robbers, the number of such birds would quickly be doubled and quadrupled.

And we must summon the law to our aid. Though law can do little or nothing against stealthy, skulking nest-plunderers, it can help us materially in our warfare with the cowardly vagabonds who traverse our fields with musket or rifle, blazing away at every unsuspecting robin or thrush that they can discover. Make it trespass, punishable with fine and imprisonment, to shoot on another's land without his express permission, and the cowardly massacre of the farmers' humble allies would be checked at once, and, when public sentiment had been properly enlightened, might, in civilized regions, be arrested altogether.

XXIII.

ABOUT TREE-PLANTING.

I have had so little experience in Tree-Planting that I should have preferred to say no more about it; but letters that have reached me imply that the ignorance of others is even denser than mine. For the sake of those only who are conscious that they know nothing, yet are not unwilling to learn, I venture a few timid suggestions with regard to Tree-Planting.

I. Ten or twelve years ago, I bought a pound or more of Locust seed rather late in the Spring, scalded it by plunging for a moment the little cotton bag which held it into a pot of boiling water, and letting the seed steep and steam in the bag till next morning, when the seed was planted in rows in a newly broken bit of poor old pasture-land. This was a mistake; I should have given that seed the richest available spot in my garden, to say nothing of planting it as early as April 20th. My locusts came up slowly and grew feebly that year, not to speak of the many seeds that did not sprout at all. Still many came up and survived, and my place is this day the richer for them.

It might have been still richer had I seasonably known more.

II. What I would now advise as to Locust and most other trees is that the best seed be procured in the Fall, or so soon as it drops from the trees ; that part of it be sown in drills, two feet apart, with two inches between seeds in the drills, and that the richest of dry, warm garden-soil be devoted to this purpose. Fill a large box with rich loam, stir four ounces of seed into this, and set the box in a cool cellar where frost does not enter, and here let it remain till April ; then take out the seed and earth together, and sow in drills as above. If some one who cuts Locust during the Winter or Spring will allow you to trace the smaller surface-roots from the new-made stumps and cut or dig them up, cut fifty or a hundred pieces of root the size of your finger each two feet long, and plant these, about May 1, in the places where you want Locusts to come forward most rapidly. Some of them may not grow, but I think many will ; and, from all these sources, I judge that you will obtain a good supply of young trees. Let those you start from the seed get two years' growth before you take them up and set them where you want trees, whether in your present woods, in rugged, rocky pastures, on the sides of steep ravines, or around your buildings. You cannot fail to obtain *some* trees if you follow these directions.

III. Begin early this Fall to gather Chestnuts, Hickory Nuts Walnuts, White Oak Acorns, etc., to plant. Select the largest and finest nuts, giving the preference to those which ripen and fall earliest. Keep them in cool, damp earth in some barn or cellar where rats and mice cannot reach them, and persist in collecting till December. Then plant a part in your garden or in any rich ground where they are not likely to be disturbed ; letting the residue remain in the boxes of moist earth where you first placed them till early Spring; then plant these, like the former, in rows two feet apart, with six inches between seed and seed in each row, and give the rows careful culture for two years; after which, set them where you wish them to grow.

I venture to suggest that he who has a rugged, stony hill or other lot which he wishes to surrender to forest should plow it, if it can be plowed, next September or October; if too rocky to be even imperfectly plowed, dig up the earth with pick and spade, and sow it thickly with hickory nuts, walnuts, chestnuts, locust and other tree-seeds, expecting that some will be dug up and carried off by squirrels, etc., and that others will fail to germinate. Go over it with hoes the ensuing June or July, killing all weeds and other infestations ; and, nearly a year later, repeat the operation, taking up young trees from your garden or nursery, and filling them in wherever there is room. Plant thickly in order to force an upward rather than a scraggy growth ; and so that you may begin to cut out the superfluous saplings for bean-poles, hoop-poles, etc., three or four years thereafter.

Cut late in Winter or early in Spring, so that the stumps will each throw up two or more shoots or sprouts, which usually grow much faster than the original tree did. And the process of thinning may thus be continued indefinitely, while the choicer trees are allowed to attain their stateliest proportions. And thus a rocky, sterile hill-side or knoll may be made to yield a crop annually after the first two or three years from planting, while growing trees of decided value. I judge that almost any land within fifty miles of a great city and not more than two miles from a railroad depot or from navigable water may thus be made to earn a good interest on $100 per acre, after meeting all the cost of breaking up and planting. I confidently assert that many thousands of sterile, rocky acres, which now yield less than $5 per acre annually in pasturage, would net at least double that sum to the owner if wisely devoted to forest-trees.

I have a hearty love of forests. They proffer gentle companionship to the thoughtful and rest to the overworked, fevered brain. Our streams will be fuller and less capricious, our gales less destructive, our climate more equable, when we shall have re-clothed our rugged slopes and rocky crests with trees. Timber grows yearly scarcer and dearer, when it ought to be becoming more plentiful and accessible, and *would* be if we devoted to trees all the land which we cultivate at a loss or fail to cultivate at all. Let our boys be incited to gather seeds and plant nurseries; let young trees be bought by the thousand where they now are by the dozen, and let us all coöperate in covering our unsightly rocks and making glad our waste places by a superabundance of choice, thrifty, healthy trees.

Many of our young men have a taste for adventure and excitement which leads them to the ocean, the mines, to Australia or some other far-off land recently and scantily peopled by civilized beings. I will not quarrel with their taste; but I judge that there are openings for their enterprise and daring within the area of our own country. Let one thousand of them resolve to devote the next five years to planting forests on the treeless plains and virtual deserts of the Great Basin and on either side of it; let them select locations where some acres may cheaply and surely be irrigated, and, having carefully provided themselves with an abundance of the best seeds, let them start patches of woodland at points the most remote from present timber, until a thousand different forests —one to each of the associates— shall have been started and guarded till their roots have taken firm hold of the earth. I presume Congress would grant them preëmptions to each section on which they thus planted at least forty acres of forest, and that most of these preëmption rights could, within ten years, be sold to settlers for many times their original cost.

XXIV.

IF I were asked to say what single aspect of our economic condition most strikingly and favorably distinguished the people of our Northern States from these of most if not all other countries which I have traversed, I would point at once to the fruit-trees which so generally diversify every little as well as larger farm throughout these States, and are quite commonly found even on the petty holdings of the poorer mechanics and workmen in every village and in the suburbs and outskirts of every city. I can recall nothing like it abroad, save in two or three of the least mountainous and most fertile districts of northern Switzerland. Italy has some approach to it in the venerable olive-trees which surround or flank many, perhaps most, of her farm-houses, upholding grape-vines as ancient and nearly as large as themselves; but the average New-England or Middle State homestead, with its ample Apple-orchard and its cluster of Pear, Cherry and Plum-trees surrounding its house and dotting or belting its garden, has an air of comfort and modest thrift which I have

(139)

nowhere else seen fairly equaled. Upland Virginia
and the mountainous portion of the States southward
of her may in time surpass the most favored regions
of the North in the abundance, variety and excellence
of their fruits ; for the Peach and the Grape find here
a congenial climate, while they are grown with difficulty, where they can be grown at all, in the North ;
but, up to this hour, I judge that our country north
of the Potomac is better supplied with wholesome
and palatable tree-fruits than any other portion of
the earth's surface of equal or nearly equal area.

On the whole, I deem it a misfortune that our
Northern States were so admirably adapted to the
Apple and kindred fruit-trees that our pioneer forefathers had little more to do than bury the seeds in
the ground and wait a few years for the resulting
fruit. The soil, formed of decayed trees and their
foliage, thickly covered with the ashes of the primitive forest, was as genial as soil could be ; while the
remaining woods, which still covered seven-eighths
of the country, shut out or softened the cold winds
of Winter and Spring, rendering it less difficult, a
century ago, to grow fine peaches in southern New-
Hampshire than it now is in southern New-York.
Devastating insects were precluded by those great,
dense woods from diffusing themselves from orchard
to orchard as they now do. Snows fell more heavily
and lay longer then than now, protecting the roots
from heavy frosts, and keeping back buds and blossoms in Spring, to the signal advantage of the husband-

man. I estimate that my apple-trees would bear at least
one-third more fruit if I could retard their blossoming a fortnight, so as to avoid the cold rains and cutting winds, often succeeded by frosts, which are apt
to pay their unwelcome farewell visits just when my
trees are in bloom or when the fruit is forming directly thereafter. Hence, I say to every one who
shall hereafter set an orchard, Give it the northward
slope of a hill if that be possible. Other things being
equal, the orchard which blossoms latest will, in a
series of years, yield most fruit, and will be most
likely to bear when the Apple-crop of your vicinity
proves a failure. I do not recommend storing ice to
plant or bury under the trees in April, for that involves too much labor and expense; yet I have no
doubt that even that has been and sometimes might
be done with profit. In the average, however, I
judge that it would not pay.

In locating and setting an orchard, the very first
consideration is thorough drainage. Nothing short
of a destructive fire can be more injurious to an
apple-tree than compelling it to stand throughout
Winter and Spring in sour, stagnant water. Barrenness, dead branches, and premature general decay,
are the natural and righteous consequences of such
crying abuse. There are many reasons for choosing
sloping or broken ground for an apple-orchard, whereof comparative exemption from frost and natural facility of drainage are the most obvious. A level field,
thoroughly undrained to-day, may, through neglect

and the mischiefs wrought by burrowing animals, have become little better than a morass thirty years hence ; but an orchard set on a tolerably steep hill-side is reasonably secure against wet feet to the close of its natural life.

A gravelly or sandy loam is generally preferred for orchards; yet I have known them to flourish and bear generously on heavy clay. Whoever has a gravelly field will wisely prefer this for Apples, not merely to clay but to sand as well.

And, while many young orchards have doubtless been injured by immoderate applications of rank, green manures, I doubt that any man has ever yet bestowed too much care and expense on the preparation of his ground for fruit-trees. Where ridges or plateaus of fast stone do not forbid, I would say, Turn over the soil to a depth of at least fifteen inches with a large plow and a strong team ; then lift and pulverize the subsoil to a depth of not less than nine inches ; apply all the Wood-ashes you can get, with one thousand bushels of Marl if you are in a Marl region ; if not, use instead from thirty to fifty bushels of quick Lime (oyster-shell if that is to be had) with one hundred loads per acre of Swamp Muck which has lain a year on dry upland, baking in the sun and wind; and now you may think of setting your trees. If your soil was rich Western prairie or Middle-State garden to begin with, you can dispense with all these fertilizers ; yet I doubt that there is an acre of Western prairie that would not be improved by the Lime or (perhaps better still) a smaller quantity of refuse Salt from a packing-house or meat retailing grocery. There are not many farms that would not repay the application of five bushels per acre of refuse Salt at twenty-five cents per bushel.

Your trees once set—(and he who sets twenty trees per day as they should be set, with each root in its natural position, and the earth pressed firmly around its trunk, but no higher than as it originally grew, is a faithful, efficient worker), I would cultivate the land, (for the trees' sake), growing crops successively of Ruta Bagas, Carrots, Beets, and early Potatoes, but no grain whatever, for six or seven years, disturbing the roots of the trees as little as may be, and guarding their trunks from tug, or trace, or whiffle-tree, by three stakes set firmly in the ground about each tree, not so near it as to preclude constant cultivation with the hoe inside as well as outside of the stakes, so as to let no weed mature in the field. Apply from year to year well-rotted compost to the field in quantity sufficient fully to counterbalance the annual abstraction by your crops. Make it a law inflexible and relentless that no animal shall be let into this orchard to forage, or for any purpose whatever but to draw on manures, to till the soil, and to draw away the crops. Thus until the first blossoms begin to appear on the trees; then lay down to grass *without* grain, unless it be a crop of Rye or Oats to be cut and carried off for feed when not more than half

grown, leaving the ground to the young grass. Let
the grass be mowed for the next two or three years, and
thenceforward devote it to the pasturage of Swine,
running over it with a scythe once or twice each Summer to clear it of weeds, and taking out the Swine a
few days before beginning to gather the Apples, but
putting them back again the day after the harvest is
completed. Let the Swine be sufficiently numerous
and hungry to eat every apple that falls within a few
hours after it is dropped, and to insure their rooting
out every grub or worm that burrows in the earth
beneath the trees, ready to spring up and apply himself to mischief at the very season when you could
best excuse his absence. I do not commend this as
all, or nearly all, that should be done in resistance to
the pest of insect ravage ; but I begin with the Hog
as the orchardist's readiest, cheapest, most effective
ally or servitor in the warfare he is doomed unceas-
ingly to wage against the spoilers of his heritage. I
will indicate some further defensive enginery in my
next chapter.

XXV.

In my opinion, Apple-trees, in most orchards. are planted too far apart and allowed to grow taller and spread their limbs more widely than is profitable. I judge that a pruner or picker should be able to reach the topmost twig of any tree with a ten-foot pole, and that no limb should be allowed to extend more than eight feet from the trunk whence it springs. Our Autumnal Equinox occurs before our Apples are generally ripe for harvest, and, finding our best trees bending under a heavy burden of fruit, its fierce gales are apt to make bad work with trees as well as apples. The best tree I had, with several others, was thus ruined by an equinoctial tempest a few years since. Barren trees escape unharmed, while those heavily laden with large fruit are wrenched and twisted into fragments. And, even apart from this peril, a hundred weight of fruit at or near the extremity of limbs which extend ten or twelve feet horizontally from the trunk, tax and strain a tree more than four times that weight growing within four or five feet of the trunk, and on limbs that maintain a semi-erect position. I diffidently sug-

7 (145)

gest, therefore, that no apple-tree be allowed to exceed fifteen feet in hight, nor to send a limb more
than eight feet from its trunk, and that trees be set
(diamond-fashion) twenty-four feet apart each way,
instead of thirty-two, as some of mine were. I judge
that the larger number of trees (72 per acre) will produce more fruit in the average than the larger but
fewer trees grown on squares of two by two rods to
each, that they will thrive and bear longer, and that
not one will be destroyed or seriously harmed by
winds where a dozen would if allowed to grow as
high and spread as far as they could.

Every apple-tree should be pruned each year of
its life : that is, it should be carefully examined with
intent to prune if that be found necessary. It should
be pruned with a careful eye to giving it the proper
shape, which, from the point where it first forks upward, should be that of a tea-cup, very nearly. I
have seen young trees so malformed that they could
rarely, if ever, bear fruit enough to render them profitable. And the pruning should be so carefully, judiciously done from the outset that no wood two years
old should ever be cut away. With old, malformed,
diseased, worm-eaten, decaying trees, the best must
be done that can be; but he who, pruning a tree that
he set and has hitherto cared for, finds himself obliged to cut off a limb thicker than his thumb, may
justly suspect himself of lacking a mastery of the art
of fruit-growing.

Sprouts from the root of an apple-tree remind me of children who habitually play truant or are kept out of school. They not merely can never come to good, but they are a nuisance to the neighborhood and bring reproach on the community.

The apple-grower should never forget that every producer needs to be fed in proportion to his product. If a cow gives twenty quarts of milk per day, she needs more grass or other food than if she gave but two quarts; and an acre of orchard that yields a hundred barrels of Apples per annum needs something given to the soil to balance the draft made upon it. Nature offers us good bargains; but she does not trust and will not be cheated. When she offers a bushel of Corn for a bushel of dirty Salt, Shell Lime, or Wood-Ashes, a load of Hay for a load of Muck, we ought not to stint the measure, but pay her demand ungrudgingly.

And now a last word on Insects.

My township (Newcastle) is said to have formerly grown more Apples per annum than any other township in the United States; its apple-trees are still as numerous as ever, but their product has fallen off deplorably. I estimate the average yield of the last three years at less than a bushel per annum for each full-grown tree; I think a majority of the trees have not borne a bushel each in all these three years. Unseasonable frosts, storms, etc., have borne the blame of this barrenness—perhaps justly, if we consider only immediate causes—but the caterpillar and other vermin are, in my view, our more potent, though remoter, afflictions. Not less than four times within the last sixteen years have our trees been covered with nests and worms; and I have seen whole orchards stripped of nearly every leaf till they were as bare (of every thing but caterpillars) in July as they should have been in December. After the scourge had passed, the trees reclad themselves with leaves; but they grew old under that visitation faster in one year than they would have done in ten of healthful fruit-bearing; and they are now prematurely gray and moss-covered because of the terrible infliction.

I lay down the general proposition that no man who harbors caterpillars has any moral right to Apples—that each grower should be required to make his choice between them. Slovenly farmers say, " O there are so many of them that I cannot kill half so fast as they multiply " Then I say, cut down and burn up the trees you can best spare, until you have no more left than you can keep clear of worms.

If it were the law of the land that whoever allowed caterpillars to nest and breed in his fruit-trees should pay a heavy fine for each nest, we should soon be comparatively clear of the scourges. In the absence of such salutary regulation, one man fights them with persistent resolution, only to see his orchard again and again invaded and ravaged by the pests hatched and harbored by his careless neighbors. He thus pays and repays the penalty of others' negligence and misdoing until, discouraged and demoralized, he abandons the hopeless struggle, and thenceforth repels the enemy from a few favorite trees around his dwelling, and surrenders his orchard to its fate. Thus bad laws (or no laws) are constantly making bad farmers. The birds that would help us to make head against our insect foes are slaughtered by reckless boys—many of them big enough to know better—and our perils and losses from enemies who would be contemptible if their numbers did not render them formidable increase from year to year. We must change all this; and the first requisite of our situation is a firm alliance of the entire farming and fruit-growing interest defensive as to birds, offensive toward their destroyers, and toward the vermin multiplied and shielded by the ruthless massacre of our feathered friends.

Since the foregoing was written we have had (in 1870) the greatest Apple-crop throughout our section that mine eyes did ever yet behold. It was so abundant that I could not sell all my cider-apples to the vinegar-makers, even at fifty cents per barrel. This establishes the continued capacity of our region to bear Apples, and should invite to the planting of new orchards and the fertilization and renovation of old ones.

XXVI.

HAY AND HAY-MAKING.

THE Grass-crop of this, as of many, if not most, other countries, is undoubtedly the most important of its annual products; requiring by far the largest area of its soil, and furnishing the principal food of its Cattle, and thus contributing essentially to the subsistence of its working animals and to the production of those Meats which form a large and constantly increasing proportion of the food of every civilized people. But I propose to speak in this essay of that proportion of the Grass-crop—say 25 to 35 per cent. of the whole—which is cut, cured and housed (or stacked) for Hay, and which is mainly fed out to animals in Winter and Spring, when frost and snow have divested the earth of herbage or rendered it inaccessible.

The Seventh Census (1850) returned the Hay-crop of the preceding year at 13,838,642 tuns, which the Eighth Census increased to 19,129,128 tuns as the product of 1859. Confident that most farmers under-estimate their Hay-crops, and that hundreds of thousands who do not consider themselves farmers, but who own or rent little homesteads of two to ten acres each, keeping thereon a cow or two and often a horse, fail to make returns of the two to five tuns of Hay they annually produce, considering them too trivial, I estimate the actual Hay-crop of all our States and Territories for the current year at 40,000,000 tuns, or about a tun to each inhabitant although I do not expect the new Census to place it much, if any, above 25,000,000 tuns. The estimated average value of this crop is $10 (gold) per tun, making its aggregate value, at my estimate of its amount, $400,000,000 — and the quantity is constantly and rapidly increasing.

That quantity should be larger from the area devoted to meadows, and the quality a great deal better. I estimate that 30,000,000 acres are annually mowed to obtain these 40,000,000 tuns of Hay, giving an average yield of 1⅓ tuns per acre, while the average should certainly not fall below two tuns per acre. My upland has a gravelly, rocky soil, not natural to grass, and had been pastured to death for at least a century before I bought it; yet it has yielded me an average of not less than 2½ tuns to the acre for the last sixteen years, and will not yield less while I am allowed to farm it. My lowland (bog when I bought it) is bound henceforth to yield more; but, while imperfectly or not at all drained, it was of course a poor reliance—yielding bounteously in spots, in others, little or nothing.

In nothing else is shiftless, slovenly farming so apt to betray itself as in the culture of Grass and the management of grass lands. Pastures overgrown with bushes and chequered by quaking, miry bogs; meadows foul with every weed, from white daisy up to the rankest brakes, with hill-sides that may once have been productive, but from which crop after crop has been taken and nothing returned to them, until their yield has shrunk to half or three-fourths of a tun of poor hay, these are the average indications of a farm nearly run out by the poorest sort of farming. Such farms were common in the New England of my boyhood; I trust they are less so to-day; yet I seldom travel ten miles in any region north or east of the Delaware without seeing one or more of them.

Fifty years ago, I judge that the greater part of the hay made in New-England was cut from sour, boggy land, that was devoted to grass simply because nothing else could be done with it. I have helped to carry the crop off on poles from considerable tracts on which oxen could not venture without miring. It were superfluous to add that no well-bred animal would eat such stuff, unless the choice were between it and absolute starvation. In many cases, a very little work done in opening the rudest surface-drains would have transformed these bogs into decent meadows, and the product, by the help of plowing or seeding, into unexceptionable hay.

There are not many farmers, apart from our wise and skillful dairymen, who use half enough grass-seed; men otherwise thrifty often fail in this respect. If half our ordinary farmers would thoroughly seed down a full third of the area they usually cultivate, and devote to the residue the time and efforts they now give to the whole, they would grow more grain and vegetables, while the additional grass would be so much clear again.

We sow almost exclusively Timothy and Clover, when there are at least 20 different grasses required by our great diversity of soils, and of these three or four might often be sown together with profit; especially in seeding down fields intended for pasture, we might advantageously use a greater variety and abundance of seed. I believe that there are grasses not yet adopted and hardly recognized by the great body of our farmers—the buffalo-grass of the prairies for one—that will yet be grown and prized over a great part of our country.

As for Hay-Making, my conviction is strong that our grass is cut in the average from two to three weeks too late, and that not only is our hay greatly damaged thereby, but our meadows needlessly impoverished and exhausted. The formation and perfection of seed always draw heavily upon the soil. A crop of grass cut when the earliest blossoms begin to drop—which, in my judgment, is the only right time—will not impoverish the soil half so much as will the same crop cut three weeks later; while the roots of the earlier cut grass will retain their vitality at least thrice as long as though half the seed had ripened before the crop was harvested. Grass that was fully ripe when cut has lost at least half its nutri-

*

ment, which no chemistry can ever restore. Hay alone is dry fodder for a long Winter, especially for young stock; but hay cut after it was dead ripe, is proper nutriment for no animal whatever—not even for old horses, who are popularly supposed to like and thrive upon it.

The fact that our farmers are too generally short-handed throughout the season of the Summer harvest, while it seems to explain the error I combat, renders it none the less disastrous and deplorable. I estimate the depreciation in the value of our hay-crop, by reason of late cutting, as not less than one-fifth; and, when we consider that a full half of our farmers turn out their cattle to ravage and poach up their fields in quest of fodder a full month earlier than they should, because their hay is nearly or quite exhausted, the consequences of this error are seen to diffuse them-selves over the whole economy of the farm.

From the hour in which grass falls under the Mower, it ought to be kept in motion until laid at rest in the stack or the barn; keep stirring it with the tedder until it is ready to be raked into light winrows, and turn these over and over until they will answer to go upon the cart. In any bright, hot day, the grass mowed in the morning should be stacked before the dew falls at night; while, if any is mowed after noon, it should be cocked and capped by sunset, even though it be necessary to open it out the next fair morning.

I have a dream of hay-making, especially with regard to clover, without allowing it to be scalded by fierce sunshine. In my dream, the grass is raked and loaded nearly as fast as cut, drawn to the barn-yard, and there pitched upon an endless apron, on which it is carried slowly through a drying-house, heated to some 200° Fahrenheit by steam or by charcoal in a furnace below, somewhat after the manner of a hop-kiln. While passing slowly through this heated atmosphere, the grass is continually forked up and shaken so as to expose every lock of it to the drying heat, until it passes off thereby deprived of its moisture and is precipitated into a mow or upon a stack-bottom at the opposite side; load after load being pitched upon the apron continuously, and the drying process going steadily forward by night as well as by day, and without regard to the weather outside. I do not assert that this vision will ever be realized; but I have known dreams as wild as this transformed by time and thought into beneficent realities.

I ask no one to share my dreams or sympathise with their drift and purpose. I only insist that Hay-making, as it is managed all around me, is ruder in its processes and more uncertain in its results than it should or need be. We cut our grass rapidly and well; we gather and house it with tolerable efficiency; but we cure much of it imperfectly and wastefully. The fact that most of it is over-ripe when cut aggravates the pernicious effects of its subsequent exposure to dew and rain; and the net result is damaged fodder which is at once unpalatable and innutritious.

XXVII.

OUR harsh, capricious climate north of the latitudes of Philadelphia, Cincinnati, and St. Louis—so much severer than that of corresponding latitudes in Europe—is unfavorable, or at least very trying, to all the more delicate and luscious Fruits, berries excepted. Except on our Pacific coast, of which the Winter temperature is at least ten degrees milder than that of the Atlantic, the finer Peaches and Grapes are grown with difficulty north of the fortieth degree of latitude, save in a few specially favored localities, whereof the southern shore of Lake Erie is most noted, though part of that of Lake Ontario and of the west coast of Lake Michigan are likewise well adapted to the Peach.

It is not the mere fact that the mercury in Fahrenheit's thermometer sometimes ranges below zero, and the earth is deeply frozen, but the suddenness wherewith such rigor succeeds and is succeeded by a temperature above the freezing point, that proves so inhospitable to the most valued Tree-Fruits. And, as the dense forests which formerly clothed the Alleghenies and the Atlantic slope, are year by year swept away, the severity of our "cold snaps," and the celerity with which they appear and disappear, are constantly aggravated. A change of 60°, or from 50° above to 10° below zero, between morning and the following midnight, soon followed by an equally rapid return to an average November temperature, often proves fatal even to hardy forest-trees. I have had the Red Cedar in my woods killed by scores during an open, capricious Winter; and my observation indicates the warmest spots in a forest as those where trees are most likely to be thus destroyed. After an Arctic night, in which they are frozen solid, a bright sun sends its rays into the warmest nooks, whence the wind is excluded, and wholly or partially thaws out the smaller trees; which are suddenly frozen solid again so soon as the sunshine is withdrawn; and this partly explains to my mind the fact that peach-buds are often killed in lower and level portions of an orchard, while they retain their vitality on the hill-side and at its crest, not 80 rods distant from those destroyed. The fact that the colder air descends into and remains in the valleys of a rolling district contributes also to the correct explanation of a phenomenon which has puzzled some observers.

Unless in a favored locality, it seems to me nnadvisable for a farmer who expects to thrive mainly by the production of Grain and Cattle, to attempt the growing of the finer Fruits, except for the use of his own family. In a majority of cases, a multiplicity of cares and labors precludes his giving to his Peaches and Grapes,. his Plums and Quinces, the seasonable and persistent attention which they absolutely require. Quite commonly, a farmer visits a grand nursery, sees with admiration its trees and vines loaded with the most luscious Fruits, and rashly infers that he has only to buy a good stock of like Trees and Vines to insure himself an abundance of delicious fruit. So he buys and sets; but with no such preparation of the soil, and no such care to keep it mellow and free from weeds, or to baffle and destroy predatory insects, as the nurseryman employs. Hence the utter disappointment of his hopes; borers, slugs, caterpillars, and every known or unknown species of insect enemies, prey upon his neglected favorites. At intervals, some domestic animal or animals get among them, and break down a dozen in an hour. So, the far greater number come to grief, without having had one fair chance to show what they could do, and the farmer jumps to the conclusion that the nurseryman was a swindler, and the trees he sells scarcely related to those whose abundant and excellent fruits tempted him to buy. I counsel every farmer to consider thoughtfully the treatment absolutely required for the production of the finer Fruits before he allows a nurseryman to make a bill against him, and not expect to grow Duchesse Pears as easily as Blackberries, or Ionas and Catawbas as readily as he does Fox-grapes on the willows which overhang his brook; for if he does he will surely be disappointed.

Some of our hardier and coarser Grapes—the Concord preëminent among them—are grown with considerable facility over a wide extent of our country; and many farmers, having planted them in congenial soil, and tended them well throughout their infancy, are rewarded by a bounteous product for two or three years. Believing their success assured, they imagine that their vines may henceforth be neglected, and in the course of two or three more years they are often utterly ruined. I know that there are wild grapes of some value, in the absence of better, which thrive and bear without attention; but I do not believe that any grape which will sell in a market where good fruit was· ever seen, can be grown north of Philadelphia but by constant care and labor, or at a cost of less than five cents per pound, under the most judicious and skillful treatment. In California, and I presume in most of our States south of the Potomac and Ohio, choice grapes may be grown more abundantly and more cheaply. Yet I think the localities are few and far between in which a tun of good grapes can be grown as cheaply as a tun of wheat, under the most judicious cultivation in either case.

I do not mean to discourage grape-growing; on the contrary, I would have every farmer, even so far north as Vermont and Wisconsin, experiment cautiously with a dozen of the most promising varieties,

including always the more hardy, in the hope of finding some one or more adapted to his soil, and capable
of enduring his climate. Even in France, the land
of the vine, one farm will produce a grape which the
very next will not: no man can satisfactorily say
why. The farmer, who has tried half a dozen grapes
and failed with all, should not be deterred from further experiments, for the very next may prove a success. I would only say, Be moderate in your expectations and careful in your experiments; and never
risk even $100 on a vineyard, till you have ascer-
tained, at a cost of $5 or under, whether the species
you are testing will thrive and bear on your soil.

In my own case, my upland mainly sloping to the
west, with a hill rising directly south of it, I have
had no luck with Grapes, and I have wasted little
time or means upon them. I have done enough to
show that they can be grown, even in such a locality,
but not to profit or satisfaction.

I would advise the farmer who proposes to grow
Pears, Peaches, and Quinces, for home use only or
mainly, to select a piece of dry, gravelly or sandy
loam, underdrain it thoroughly, plow or trench it
very deeply, and fertilize it generously, in good part
with ashes and with leaf-mold from his woods. Locate the pig-pen on one side of it, fence it strongly,
and let the pigs have the run of it for a good portion
of each year. In this plat or yard, plant half a dozen
Cherry and as many Pear trees of choice varieties,
the Bartlett foremost among them ; keep clear of all

dwarfs, and let your choicest trees have a chance to run under the pig-pen if they will.   Plant here also, if your climate does not forbid, a dozen well-chosen Peach-trees, and two each year thereafter to replace those that will soon be dying out; and give half a dozen Quinces moist and rich locations by the side of your fences; surrounding each tree with stakes or pickets that will preclude too great familiarity on the part of the swine, and will not prevent a sharp scrutiny for borers in their season.   Do not forget that a fruit-tree is like a cow tied to an immovable stake, from which you cannot continue to draw a pail of milk per day unless you carry her a liberal supply of food; and every Fall cart in half a dozen loads of muck from some convenient swamp or pond for your pigs to turn over.   Should they leave any weeds, cut them with a scythe as often as they seem to need it; never allowing one to ripen seed.   There may be easier and surer ways to obtain choice fruits; but this one commends itself to my judgment as not surpassed by any other.   I think few have grown fruits to profit but those who make this a specialty; and I feel that disappointment in fruit-culture is by no means near the end.   You *can* grow Plums, or Grapes, or Peaches, outside of the climate most congenial to them, but this is a work wherein success is likely to cost more than its worth.   Try it first on a small scale, if you will try it; and be sure you do it thoroughly.

XXVIII.

GRAIN-GROWING——EAST AND WEST.

I DISCLAIM all pretensions to ability to teach Western farmers how to grow Indian Corn abundantly and profitably, while I cheerfully admit that they have taught *me* somewhat thoroughly worth knowing. In my boyhood, I hoed Corn diligently for weeks at a time, drawing the earth from between the rows up about the stalks to a depth of three or four inches; thus forming hills which the West has since taught me to be of no use, but rather a detriment, embarrassing the efforts of the growing, hungry plants to throw out their roots extensively in every direction, and subjecting them to needless injury from drouth. I am thoroughly convinced that Corn, properly planted, will, like Wheat and all other grains, root itself just deep enough in the ground, and that to keep down all weeds and leave the surface of the cornfield open, mellow and perfectly flat, is the best as well as the cheapest way to cultivate Corn. And I do not believe that so much human food, with so little labor, is produced elsewhere on

earth as in the spacious fields of Wheat and Corn in
our grand Mississippi valley.

And yet I have seen in that valley many ample
stretches covered with Corn, whereof the tillage
seemed susceptible of improvement. Riding between
these great corn-fields in October, after everything
standing thereon had been killed by frost, it seemed
to my observation that, while the corn-crop was fair,
the weed-crop was far more luxuriant; so that, if
everything had been cut clean from the ground, and
the corn and the weeds placed in opposite scales, the
latter would have weighed down the former. I cannot doubt that the cultivation, or lack of cultivation,
which produces or permits such results, is not merely
slovenly, but unthrifty.

The West is for the present, as for a generation
she has been, the granary of the East. In my judgment, she will not long be content to remain so.
Fifty years ago, the Genesee valley supplied most of
the wheat and flour imported into New-England; ten
years later, Northern Ohio was our principal re-
source; ten years later still, Michigan, Indiana,
northern Illinois, and eastern Wisconsin, had been
added to our grain-growing territory. Another de-
cade, and our flour manufacturers had crossed the
Mississippi, laying Iowa and Minnesota under liberal
contributions, while western New-York had ceased
to grow even her own breadstuffs, and Ohio to produce one bushel more than she needed for home consumption. Can we doubt that this steady recession

of our Egypt, our Hungary, is destined to continue? Twenty-three years ago, when I first rode out from the then rising village of Chicago to see the Illinois prairies, nearly every wagon I met was loaded with wheat, going into Chicago, to be sold for about fifty cents per bushel, and the proceeds loaded back in the form of lumber, groceries, and almost everything else, grain excepted, needed by the pioneers, then dotting, thinly and irregularly, that whole region with their cabins. Now, I presume the district I then traversed produces hardly more grain than it consumes; taking Illinois altogether, I doubt that she will grow her own breadstuffs after 1880; not that she will be unable to produce a large surplus, but that her farmers will have decided that they can use their lands otherwise to greater advantage. Iowa and Minnesota will continue to export grain for perhaps twenty years longer; but even their time will come for saying, "New-York and New-England (not to speak of *Old* England) are too far away to furnish profitable markets for such bulky products; the cost of transportation absorbs the larger part of the cargo. We must export instead Wool, Meat, Lard, Butter, Cheese, Hops, and various Manufactures, whereof the freight will range from 2 up to not more than 25 per cent. of the value." They will thus save their soil from the tremendous exaction made by taking grain-crop after grain-crop persistently, which long ago exhausted most of New-England and eastern New-York cf wheat-forming material, and has since

wrought the same deplorable result in our rich Gen-
esce valley; while eastern Pennsylvania, though settled nearly two centuries ago, having pursued a more
rational and provident system of husbandry, grows
excellent wheat-crops to this day.

I insist that the States this side of the Delaware,
though they will draw much grain from the Canadas
after the political change that cannot be far distant,
will be compelled to grow a very considerable share
of their own breadstuffs; that the West will cease to
supply them unless at prices which they will deem exorbitant; and that grain-growing eastward of a line
drawn from Baltimore due north to the Lakes will
have to be very considerably extended. Let us see,
then, whether this might not be done with profit even
now, and whether the East is not unwise in having so
generally abandoned grain-growing.

I leave out of the account most of New-England, as
well as of Eastern New-York, and the more rugged
portions of New-Jersey and Pennsylvania, where the
rocky, hilly, swampy face of the country seems to
forbid any but that *patchy* cultivation, wherein
machinery and mechanical power can scarcely be
made available, and which seem, therefore, permanently fated to persevere in a system of agriculture
and horticulture not essentially unlike that they now
exhibit. In the valleys of the Penobscot, the Kennebec, the Hudson, and of our smaller rivers, there
are considerable tracts absolutely free from these
natural impediments, whereon a larger and more

efficient husbandry is perfectly practicable, even now; but these intervales are generally the property of many owners; are cut up by roads and fences; and are held at high prices: so that I will simply pass them by, and take for illustration the "Pine Barrens" of Southern New-Jersey, merely observing that what I say of them is equally applicable, with slight modifications, to large portions of Long Island, Delaware, Maryland, Virginia, and the Carolinas.

The "Pine Barrens" of New-Jersey are a marine deposit of several hundred feet in depth, mainly sand, with which more or less clay is generally intermingled, while there are beds and even broader stretches of this material nearly or quite pure; the clay sometimes underlying the sand at a depth of 10 to 30 or 40 inches. Vast deposits of muck or leaf-mold, often of many acres in extent and from two to twenty feet in depth, are very common; so that hardly any portion of the dry or sandy land is two miles distant from one or more of them, while some is usually much nearer; and half the entire region is underlaid by at least one stratum of the famous marl (formed of the decomposed bones of gigantic marine monsters long ago extinct) which has already played so important and beneficent a part in the renovation and fertilization of large districts in Monmouth, Burlington, Salem, and other counties.

Let us suppose now that a farmer of ample means and generous capacity should purchase four hundred acres of these "barrens," with intent to produce therefrom, not sweet potatoes, melons, and the "truck" to which Southern Jersey is so largely devoted, but substantial Grain and Meat; and let us see whether the enterprise would probably pay.

Let us not stint the outlay, but, presuming the tract to be eligibly located on a railroad not too distant from some good marl-bed, estimate as follows:

Purchase-money of 400 acres at $25 per acre............$10,000
Clearing, grubbing, fencing and breaking up ditto at $20
    per acre, over and above the proceeds of the wood..... 8,000
One thousand bushels of best Marl per acre, at 6 cents per
    bushel delivered..................................... 24,000
One hundred loads of Swamp Muck, per acre, at 50 cents
    per load.. 20,000
Fifty bushels (unslaked) of Oyster-shell Lime (to compost
    with the Muck), per acre, at 25 cents per bushel, deliv-
    ered.. 5,000
One hundred tuns of Bone Flour at $50 per tun......... 5,000

[Net cost, $180 per acre.]   Total.............$72,000

I believe that this tract, divided by light fences into four fields of 100 acres each, and seeded in rotation to Corn, Wheat, Clover and other grasses, would produce fully 60 bushels of Corn and 30 of Wheat per acre, with not less than 3 tuns of good Hay; and that by cutting, steaming, and feeding the stalks and straw on the place, not pasturing, but keeping up the stock, and feeding them, as indicated in a former chapter of these essays, and selling their product in the form of Milk, Butter, Cheese and Meat, a greater profit would be realized than could be from a like invest-

ment in Iowa or Kansas. The soil is warm, readily
frees itself, or is freed, from surplus water; is not
addicted to weeds; may be plowed at least 200 days
in a year; may be sowed or planted in the Spring,
when Minnesota is yet solidly frozen; while the crop,
early matured, is on hand to take advantage of any
sudden advance in the European or our own seaboard
markets. Labor, also, is cheaper and more rapidly
procured in the neighborhood of this great focus of
immigration than it is or can be in the West; and
our capable farmers may take their pick of the workers thronging hither from Europe, at the moment of
their landing on our shore. Of course, the owner of
such an estate as I have roughly outlined, would be
likely to keep a part of his purchase in timber, improving the quality thereof by cutting out the less
desirable. trees, trimming up the rest, and planting
new ones among them; and he would be almost certain to devote some part of his farm annually to the
growth of Roots, Vegetables, and Fruits. But I have
aimed to show only that he would grow grain here
at a profit, and I think I have succeeded. His 60
bushels of corn (shelled) per acre could be sold at his
crib, one year with another, for 60 silver dollars; and
he need seldom wait a mouth after husking it for
customers who would gladly take his grain and pay
the money for it. This would be just about double
what the Iowa or Missouri farmer can expect to
average for *his* Corn. The abundant fodder would
also be worth in New-Jersey at least double its value

in Iowa; and I judge that the farmer able to buy, prepare, fertilize, and cultivate 1,200 acres of the Jersey "barrens," could make more than thrice the profit to be realized by the owner of 400 acres. He would plow and seed as well as thrash, shell, cut stalks and straw, and prepare the food of his animals, wholly by steam-power, and would soon learn to cultivate a square mile at no greater expense than is now involved in the as perfect tillage of 200 acres.

This essay is not intended to prove that Grain is not or may not be profitably cultivated at the West, nor that it is unadvisable for Eastern farmers to migrate thither in order so to cultivate it. What I maintain is, that Wheat, Indian Corn, and nearly all our great food staples, may also be profitably produced on the seaboard, and that thousands of square miles, now nearly or quite unproductive, may be wisely and profitably devoted to such production. Let us regard, therefore, without alarm, the prospect of such a development and diversification of Western Industry as will render necessary a large and permanent extension (or.rather revival) of Eastern graingrowing.

8

XXIX.

In no other form can so large an amount and value of human food be obtained from an acre of ground as in that of edible roots or tubers; and of these the Potato is by far the most acceptable, and in most general use. Our ancestors, it is settled, were destitute and ignorant of the Potato prior to the discovery of America, though Europe would now find it difficult to subsist her teeming millions without it. In travelling pretty widely over that continent, I cannot remember that I found any considerable district in which the Potato was not cultivated, though Ireland, western England, and northern Switzerland, with a small portion of northern Italy, are impressed on my mind as the most addicted to the growth of this esculent. Other roots are eaten occasionally, by way of variety, or as giving a relish to ordinary food; but the Potato alone forms part of the every-day diet alike of prince and peasant. It is an almost indispensable ingredient of the feasts of Dives, while it is the cheapest and commonest resort for satiating or moderating the hunger of Lazarus. I recollect hearing my parents, fifty years ago, relate how, in their childhood and youth, the poor of New-England, when the grain-crop of that region was cut short, as it often was, were obliged to subsist through the following Winter mainly on Potatoes and Milk; and I then accorded to those unfortunates of the preceding generation a sympathy which I should now considerably abate, provided the Potatoes were of good quality. Roasted Potatoes, seasoned with salt and butter and washed down with bounteous draughts of fresh buttermilk, used in those days to be the regular supper served up in farmers' homes after a churning of cream into butter; and I have since eaten costly suppers that were not half so good.

The Potato, say some accredited accounts, was first brought to Europe from Virginia, by Sir Walter Raleigh in 1586 or 1587; but I do not believe the story. Authentic tradition affirms that the Potato was utterly unknown in New-England, or at all events east of the Connecticut, when the Scotch-Irish who first settled Londonderry, N. H., came over from *old* Londonderry, Ireland, bringing the Potato with them. They spent the Winter of 1719 in different parts of Massachusetts and Maine—quite a number of them at Haverhill, Mass., where they gave away a few Potatoes for seed, on leaving for their own chosen location in the Spring; and they afterward learned that the English colonists, who received them, tried hard to find or make the seed-balls edible the next Fall, but were obliged to give it up as

a bad job, leaving the tubers untouched and unsus-
pected in the ground.

I doubt that the Potato was found growing by
Europeans in any part of this country, unless it be in
that we have acquired from Mexico. It is essentially
a child of the mountains, and I presume it grew wild
nowhere else than on the sides of the great chain
which traversed Spanish America, at a height of from
5,000 to 8,000 feet above the surface of the ocean.
Here it found a climate cooled by the elevation and
moistened by melting snows from above and by frequent showers, yet one which seldom allowed the
ground to be frozen to any considerable depth, while
the pure and bracing atmosphere was congenial to its
nature and requirements. In this country, the Potato
is hardiest and thriftiest among the White Mountains
of New-Hampshire, the Green Mountains of Vermont,
on the Catskills and kindred elevations in our own
State, and in similar regions of Pennsylvania and the
States further South and West.

My own place is at least 15 miles from, and 500 feet
above, Long Island Sound; yet I cannot make the
Potato, by the most generous treatment, so prolific
as it was in New-Hampshire in my boyhood, where I
dug a bushel from 14 hills, grown on rough, hard
ground, but which, having just been .cleared of a
thick growth of bushes and briars, was probably
better adapted to this crop than though it had been
covered an inch deep with barn-yard manure.

He who has a tolerably dry, warm, or sandy soil,

covered two or three inches deep with decayed or decaying leaves and brush, may count with confidence on raising from it a good crop of Potatoes, provided his seed be sound and healthy. On the other hand, all authorities agree that animal manures, unless very thoroughly rotted and intimately mixed with the soil, are injurious to the quality of Potatoes grown thereon, stimulating any tendency to disease, if they do not originally produce such disease. I believe that Swamp Muck, dug in Summer or Autumn, deposited on a dry bank or glade, and cured of its acidity by an admixture of Wood-Ashes, of Lime, or of Salt (better still, of Lime and Salt chemically compounded by dissolving the Salt in the least possible quantity of Water, and slaking the lime with that Water), forms an excellent fertilizer for Potatoes, if administered with a liberal hand. A bushel of either of these alkalies to a cord of muck is too little; the dose should be doubled if possible; but, if the quantity be small, mix it more carefully, and give it all the time you can wherein to operate upon the muck before applying the mixture to your fields.

Where the muck is not easily to be had, yet the soil is thin and poor, I would place considerable reliance on deep plowing and subsoiling in the Fall, and cross-plowing just before planting in the Spring. Give a good dressing of Plaster, not less than 200 lbs. to the acre, directly after the Fall plowing; if you have Ashes, scatter them liberally in the drill or hill as you plant; and, if you have them not, supply their place with Super-phosphate or Bone-dust. I think many farmers will be agreeably surprised by the additional yield which will accrue from this treatment of their soil.

Those who have no swamp muck, and feel that they can afford the outlay, may, by plowing or subsoiling early in the Fall, seeding heavily with rye, and turning this under when the time comes for planting in the Spring, improve both crop and soil materially. But even to these I would say: Apply the Gypsum in the Fall, and the Ashes or Lime and Salt mixture in the Spring; and now, with good seed and good luck, you will be reasonably sure of a bounteous harvest. If a farmer, having a poor worn-out field of sandy loam, wants to do his very best by it, let him plow, subsoil, sow rye and plaster in the Fall, as above indicated, turn this under, and sow buckwheat late in the next Spring; plow this under in turn when it has attained its growth, and sow to clover; turn this down the following Spring, and plant to late potatoes, and he will not merely obtain a large crop, but have his land in admirable condition for whatever may follow.

I am quite well aware that such an outlay of labor and seed, with an entire loss of crop for one season, will seem to many too costly. I do not advise it except under peculiar circumstances; and yet I am confident that there are many fields that would be doubled in value by such treatment, which would richly repay all its cost. That most farmers could not afford thus to treat their entire farms at once, is very true ; yet it does not follow that they might not deal with field after field thus thoroughly, living on the products of 40 or 50 acres, while they devoted five or six annually to the work of thorough renovation.

A quarter of a century ago, we were threatened with a complete extinction of the Potato, as an article of food : the stalks, when approaching or just attaining maturity, were suddenly smitten with fatal disease—usually, after a warm rain followed by scalding sunshine—the growing tubers were speedily affected ; they rotted in the ground, and they rotted nearly as badly if dug ; and whole townships could hardly show a bushel of sound Potatoes.

A desolating famine in Ireland, which swept away or drove into exile nearly two millions of her people, was the most striking and memorable result of this wide-spread disaster. For several succeeding seasons, the Potato was similarly, though not so extensively, affected; and the fears widely expressed that the day of its usefulness was over, seemed to have ample justification. Speaking generally, the Potato has never since been so hardy or prolific as it was half a century ago; it has gradually recovered, however, from its low estate, and, though the malady still lingers, and from time to time renews its ravages in different localities, the farmer now plants judiciously and on fit ground, with a reasonable hope that his labor will be duly rewarded.

It seems to be generally agreed that clayey soils are not adapted to its growth ; that, if the quantity of the crop be not stinted, its quality is pretty sure to be inferior ; and I can personally testify that the planting of Potatoes on wet soil—that is, on swampy or spongy land which has not been thoroughly drained and sweetened—is a hopeless, thriftless labor— that the crop will seldom be worth the seed.

As to the ten or a dozen different insects to which the Potato-rot has been attributed, I regard them all as consequences, not causes ; attracted to prey on the plant by its sickly, weakly condition, and not really responsible for that condition. If any care for my reasons, let him refer to what I have said of the Wheat-plant and its insect enemies.*

There has been much discussion as to the kind of seed to be planted; and I think the result has been a pretty general conviction that it is better to cut the tuber into pieces having two or three eyes each, than to plant it whole, since the whole Potato sends up a superfluity of stalks, with a like effect on the crop to that of putting six or eight kernels of corn in each hill.

Small Potatoes are immature, unripe, and of course should never be planted, since their progeny will be feeble and sickly. Select for seed none but thoroughly ripe Potatoes, and the larger the better.

My own judgment favors planting in drills rather than hills, with ample space for working between them ; not less than 30 inches : the seed being drop-

* See Chapter XXII.

ped about 6 inches apart in the drill. The soil must be deep and mellow, for the Potato suffers from drouth much sooner than Indian Corn or almost any other crop usually grown among us. I believe in covering the seed from 2 to 2½ inches; and I hold to flat or level culture for this as for everything else. Planting on a ridge made by turning two furrows together may be advisable where the land is wet; but then wet land never can be made fit for cultiva-tion, except by underdraining. And I insist upon setting the rows or drills well apart, because I hold that the soil should often be loosened and stirred to a good depth with the subsoil plow; and that this process should be persevered in till the plant is in blossom. Hardly any plant will pay better for per-sistent cultivation than the Potato.

As to varieties, I will only say that planting the tubers for seed is an unnatural process, which tends and must tend to degeneracy. The new varieties now most prized will certainly run out in the course of twenty or thirty years at furthest, and must be re-placed from time to time by still newer, grown from the seed. This creation of new species is, and must be, a slow, expensive process; since not one in a hundred of these varieties possess any value. I do n't quite believe in selling—I mean in buying—Potatoes at $1 per pound; but he who originates a really valuable new Potato deserves a recompense for his in-dustry, patience, and good fortune; and I shall be glad to learn that he receives it.

8*

XXX.

ROOTS····TURNIPS—BEETS—CARROTS.

IF there be any who still hold that this country must ultimately rival that magnificent Turnip-culture which has so largely transformed the agricultural industry of England and Scotland, while signally and beneficently increasing its annual product, I judge that time will prove them mistaken. The striking diversity of climate between the opposite coasts of the Atlantic forbids the realization of their hopes. The British Isles, with a considerable portion of the adjacent coast of Continental Europe, have a climate so modified by the Gulf Stream and the ocean that their Summers are usually moist and cool, their Autumns still more so, and their Winters rarely so cold as to freeze the earth considerably; while our Summers and Autumns are comparatively hot and dry; our Winters in part intensely cold, so as to freeze the earth solid for a foot or more. Hence, every variety of turnip is exposed here in its tenderer stages to the ravages of every devouring insect; while the 1st of December often finds the soil of all but our Southern and Pacific States so frozen that cannon-wheels would hardly track it, and roots not previously dug up must remain fast in the earth for weeks and often for months. Hence, the turnip can never grow so luxuriantly, nor be counted on with such certainty, here as in Great Britain ; nor can animals be fed on it in Winter, except at the heavy cost of pulling or digging, cutting off the tops and carefully housing in Autumn, and then slicing and feeding out in Winter. It is manifest that turnips thus handled, however economically, cannot compete with hay and corn-fodder in our Eastern and Middle States; nor with these and the cheaper species of grain in the West, as the daily Winter food of cattle.

Still, I hold that our stock-growing farmers profitably may, and ultimately will, grow *some* turnips to be fed out to their growing and working animals. A good meal of turnips given twice a week, if not oftener, to these, will agreeably and usefully break the monotony of living exclusively on dry fodder, and will give a relish to their hay or cut stalks and straw, which cannot fail to tell upon their appetite, growth and thrift. Let our cattle-breeders begin with growing an acre or two each of Swedes per annum, so as to give their stock a good feed of them, sliced thin in an effective machine, at least once in each week, and I feel confident that they will continue to grow turnips, and will grow more and more of them throughout future years.

The Beet seems to me better adapted to our climate, especially south of the fortieth degree of north latitude, than any variety of the Turnip with which I am acquainted, and destined, in the good time coming, when we shall have at least doubled the average depth of our soil, to very extensive cultivation among us. I am not regarding either of these roots with reference to its use as human food, since our farmers generally understand that use at least as well as I do; nor will I here consider at length the use of the Beet in the production of Sugar. I value that use highly, believing that millions of the poorer classes throughout Europe have been enabled to enjoy Sugar through its manufacture from the Beet who would rarely or never have tasted that luxury in the absence of this manufacture. The people of Europe thus made familiar with Sugar can hardly be fewer than 100,000,000; and the number is annually increasing. The cost of Sugar to these is considerably less in money, while immeasurably less in labor, than it would or could have been had the tropical Cane been still regarded as the only plant available for the production of Sugar.

But the West Indies, wherein the Cane flourishes luxuriantly and renews itself perennially, lie at our doors. They look to us for most of their daily bread, and for many other necessaries of life; while several, if not all of them, are manifestly destined, in the natural progress of events, to invoke the protection of our flag. I do not, therefore, feel confident that Beet Sugar now promises to become an important staple destined to take a high rank among the products of our national industry. With cheap labor, I believe it might to-day be manufactured with profit in the rich, deep valleys of California, and perhaps in those of Utah and Colorado as well. On the whole, however, I cannot deem the prospect encouraging for the American promoters of the manufacture of Beet Sugar.

But when we shall have deepened essentially the soil of our arable acres, fertilized it abundantly, and cured it by faithful cultivation of its vicious addiction to weed-growing, I believe we shall devote millions of those acres to the growth of Beets for cattle-food, and, having learned how to harvest as well as till them mainly by machinery, with little help from hand labor, we shall produce them with eminent profit and satisfaction to the grower. On soil fully two feet deep, thoroughly underdrained and amply fertilized, I believe we shall often produce one thousand bushels of Beets to the acre ; and so much acceptable and valuable food for cattle can hardly be obtained from an acre in any other form.

So with regard to Carrots. I have never achieved eminent success in growing these, nor Beets ; mainly because the soil on which I attempted to grow them was not adapted to, or rather not yet in condition for, such culture. But, should I live a few years longer, until my reclaimed swamp shall have become thoroughly sweetened and civilized, I mean to grow on some part thereof 1,000 bushels of Carrots per acre, and a still larger product of Beets ; and the Carrot, in my judgment, ought now to be extensively grown in the South and West, as well as in this section, for feeding to horses. I hold that 60 bushels of Carrots and 50 of Oats, fed in alternate meals, are of at least equal value as horse-feed with 100 bushels of Oats alone, while more easily grown in this climate. The Oat-crop makes heavy drafts upon the soil, while our hot Summers are not congenial to its thrift or perfection. Since we must grow Oats, we must be content to import new seed every 10 or 15 years from Scotland, Norway, and other countries which have cooler, moister Summers than our own ; for the Oat will inevitably degenerate under such suns as blazed through the latter half of our recent June. Believing that the Carrot may profitably replace at least half the Oats now grown in this country, I look forward with confidence to its more and more extensive cultivation.

The advantage of feeding Roots to stock is not to be measured and bounded by their essential value. Beasts, like men, require a variety of food, and thrive best upon a regimen which involves a change of diet. Admit that Hay is their cheapest Winter food ; still, an occasional meal of something more succulent will prove beneficial, and this is best afforded by Roots.

XXXI.

THE FARMER'S CALLING.

IF any one fancies that he ever heard *me* flattering
farmers as a class, or saying anything which implied
that they were more virtuous, upright, unselfish, or
deserving, than other people, I am sure he must have
misunderstood or that he now misrecollects me. I
do not even join in the cant, which speaks of farmers
as supporting everybody else—of farming as the only
indispensable vocation. You may say if you will that
mankind could not subsist if there were no tillers of
the soil; but the same is true of house-builders, and
of some other classes. A thoroughly good farmer is
a useful, valuable citizen : so is a good merchant,
doctor, or lawyer. It is not essential to the true
nobility and genuine worth of the farmer's calling
that any other should be assailed or disparaged.

Still, if one of my three sons had been spared to
attain manhood, I should have advised him to try to
make himself a good farmer; and this without any
romantic or poetic notions of Agriculture as a pursuit. I know well, from personal though youthful
experience, that the farmer's life is one of labor,

anxiety, and care; that hail, and flood, and hurricane, and untimely frosts, over which he can exert no control, will often destroy in an hour the net results of months of his persistent, well-directed toil; that disease will sometimes sweep away his animals, in spite of the most judicious treatment, the most thoughtful providence, on his part; and that insects, blight, and rust, will often blast his well-grounded hopes of a generous harvest, when they seem on the very point of realization. I know that he is necessarily exposed, more than most other men, to the caprices and inclemencies of weather and climate; and that, if he begins responsible life without other means than those he finds in his own clear head and strong arms, with those of his helpmeet, he must expect to struggle through years of poverty, frugality, and resolute, persistent, industry, before he can reasonably hope to attain a position of independence, comfort, and comparative leisure. I know that much of his work is rugged, and some of it absolutely repulsive; I know that he will seem, even with unbroken good fortune, to be making money much more slowly than his neighbor, the merchant, the broker, or eloquent lawyer, who fills the general eye while he prospers, and, when he fails, sinks out of sight and is soon forgotten; and yet, I should have advised my sons to choose farming as their vocation, for these among other reasons:

I. There is no other business in which success is so nearly certain as in this. Of one hundred men who embark in trade, a careful observer reports that ninety-five fail; and, while I think this proportion too large, I am sure that a large majority do, and must fail, because competition is so eager and traffic so enormously overdone. If ten men endeavor to support their families by merchandise in a township which affords adequate business for but three, it is certain that a majority must fail, no matter how judicious their management or how frugal their living. But you may double the number of farmers in any agricultural county I ever traversed, without necessarily dooming one to failure, or even abridging his gains. If half the traders and professional men in this country were to betake themselves to farming to-morrow, they would not render that pursuit one whit less profitable, while they would largely increase the comfort and wealth of the entire community; and, while a good merchant, lawyer, or doctor, may be starved out of any township, simply because the work he could do well is already confided to others, I never yet heard of a temperate, industrious, intelligent, frugal, and energetic farmer who failed to make a living, or who, unless prostrated by disease or disabled by casualty, was precluded from securing a modest independence before age and decrepitude divested him of the ability to labor.

II. I regard farming as that vocation which conduces most directly and palpably to a reverence for Honesty and Truth. The young lawyer is often constrained, or at least tempted, by his necessities, to do the dirty professional work of a rascal intent on cheating his neighbor out of his righteous dues. The young doctor may be likewise incited to resort to a quackery he despises in order to secure instant bread; the unknown author is often impelled to write what will sell rather than what the public ought to buy; but the young farmer, acting *as* a farmer, must realize that his success depends upon his absolute verity and integrity. He deals directly with Nature, which never was and never will be cheated. He has no temptation to sow beach sand for plaster, dock-seed for clover, or stoop to any trick or juggle whatever. " Whatsoever a man soweth that shall he also reap," while true, in the long run, of all men, is instantly and palpably true as to him. When he, having grown his crop, shall attempt to sell it—in other words, when he ceases to be a farmer and becomes a trader—he may possibly be tempted into one of the many devious ways of rascality; but, so long as he is acting simply as a farmer, he can hardly be lured from the broad, straight highway of integrity and righteousness.

III. The farmer's calling seems to me that most conducive to thorough manliness of character. Nobody expects him to cringe, or smirk, or curry favor, in order to sell his produce. No merchant refuses to buy it because his politics are detested or his religious opinions heterodox. He may be a Mormon, a Rebel, a Millerite, or a Communist, yet his Grain or his Pork will sell for exactly what it is worth—not a fraction less or more than the price commanded by the kindred product of like quality and intrinsic value of his neighbor, whose opinions on all points are faultlessly orthodox and popular. On the other hand, the merchant, the lawyer, the doctor, especially if young and still struggling dubiously for a position, are continually tempted to sacrifice or suppress their profoundest convictions in deference to the vehement and often irrational prepossessions of the community, whose favor is to them the breath of life. "She will find that *that* won't go down here," was the comment of an old woman on a Mississippi steamboat, when told that the plain, deaf stranger, who seemed the focus of general interest, was Miss Martineau, the celebrated Unitarian ; and in so saying she gave expression to a feeling which pervades and governs many if not most communities. I doubt whether the social intolerance of adverse opinions is more vehement anywhere else than throughout the larger portion of our own country. I have repeatedly been stung by the receipt of letters gravely informing me that my course and views on a current topic were adverse to public opinion : the writers evidently assuming, as a matter of course, that I was a mere jumping-jack, who only needed to know what other people thought to insure my instant and abject conformity to their prejudices. Very often, in other days, I was favored with letters from indignant subscribers, who, dissenting from my views on some question, took this method of informing me that they

should no longer take my journal—a superfluous
trouble, which could only have meant dictation or
insult, since they had only to refrain from renewing
their subscriptions, and their *Tribune* would stop
coming, whenever they should have received what
we owed them ; and it would in no case stop till
then. That a journalist was in any sense a public
teacher—that he necessarily had convictions, and was
not likely to suppress them because they were not
shared by others—in short, that his calling was other
and higher than that of a waiter at a restaurant, expected to furnish whatever was called for, so long as
the pay was forthcoming—these ex-subscribers had
evidently not for one moment suspected. That such
persons have little or no capacity to insult, is very
true ; and yet, a man is somewhat degraded in his
own regard by learning that his vocation is held in
such low esteem by others. The true farmer is
proudly aware that it is quite otherwise with *his* pursuit—that no one expects him to swallow any creed,
support any party, or defer to any prejudice, as a
condition precedent to the sale of his products.
Hence, I feel that it is easier and more natural in his
pursuit than in any other for a man to work for a
living, and aspire to success and consideration, without sacrificing self-respect, compromising integrity,
or ceasing to be essentially and thoroughly a gentle-
man.

XXXII.

THE current season is quite commonly characterized as the coldest, the hottest, the wettest, or the dryest, that was ever known. Men undoubtingly assert that they never knew a Summer so hot, or a Winter so cold, when in fact several such have occurred within the cycle of their experience. Hardly anything else is so easily or so speedily forgotten as extremes of temperature or inclemencies of weather, after they have passed away. I presume there have been six to ten Summers, since the beginning of this century, as hot and as dry as that of 1870; yet the fact remains that, ·throughout the Eastern section of our country, to say nothing of the rest, the heat and drouth of the current Summer have been quite remarkable. For two months past, counting from the 10th of June, nearly every day has been a hot one, with blazing sunshine throughout, rarely interrupted and slightly modified by infrequent and inadequate showers; and, as a general result of this tropical fervor, the earth is parched and baked from ten to forty inches from the surface;

streams and ponds are dried up or shrunk to their lowest dimensions; forests are often ravaged and desolated by fires; our pastures are dry and brown; while crops of Hay, Oats, Potatoes, Buckwheat, etc., either have proved, or certainly must prove, a disappointment to the hopes of the growers. I estimate the average product for 1870 of the farms of New-England, eastern New-York and New-Jersey, as not more than two-thirds of a full harvest; while the earth remains at this moment so baked and incrusted that several days' rain is needed to fit it for Fall plowing and the sowing of Winter grain.

Such seasons must not be regarded as extraordinary. The Summer of 1854 was nearly or quite as dry as this; and I presume one or two such have intervened since that time. The heat of 1870 is remarkable for its persistence rather than its intensity. Every Summer has its heated term; that of 1870 has been longer in this region than any before it that I can remember, though doubtless the recollection of others might supply its perfect counterpart. Nearly every Summer has its drouth; the present is peculiar rather for its early commencement than its extreme duration. As our country is more and more denuded of its primitive forests, drouths longer and severer even than this may naturally be expected. What our farmers have to do is, to prepare for and provide against them.

Such seasons are disastrous to those only who farm as if none such were to be expected. Those who plow deeply, fertilize bountifully, and cultivate thoroughly, need not fear them, as fields of Hay and Oats already harvested, and of Corn and Potatoes now hastening to maturity in almost every township of the suffering region, abundantly attest. I doubt that more luxuriant crops of Corn, Tobacco, or Onions, were ever grown on the bottom-lands of the Connecticut Valley than may be seen there to-day, with failures all about them, and under drouth so fierce that Blackberries and Whortleberries are withered when half grown; even the bushes in some cases perishing for lack of moisture.

My last trip took me along the banks of the upper Hudson, through the rugged county of Warren, N. Y. The narrow, irregular intervale of this mountain stream appear to have been cultivated for the last fifty or sixty years by a hardy race, who look mainly to the timber of the wild region north of them for a subsistence. In such a district, whatever ministers to the sustenance of man or beast bears a high price; and Corn, Rye, Oats, Buckwheat, Apples and Grass, are grown wherever the soil is not too rugged or too sterile for culture. I presume half a crop of Hay has been secured throughout this valley, with perhaps a full crop of Rye where Rye was sown; but of Oats the yield will be considerably less than that, while of Corn and Buckwheat it will range from ten bushels per acre down to nothing. When I, last Summer, passed through spacious field after field of Corn in Virginia that would not mature a single ear, I spoke of it as something unknown at the North; but there are fields planted to Corn, in the upper valley of the Hudson, that will not produce a single sound ear, nor one bushel even of the shortest and poorest "nubbins;" and alongside of these are acres of Buckwheat, blossoming at an average hight of four inches, and not likely to get two inches higher.

Now, if this land were so poor or so rocky that good crops could not be extracted from it, far be it from me to disparage the agriculture whereof the results are so meagre; but I am speaking of a river intervale of considerable natural fertility, from which deep and thorough cultivation would insure ample harvests, subject only to the contingency of early frosts in Autumn. Were these lands fertilized and cultivated as they might be, and as mine are, they would yield 30 bushels of Rye or 60 of Indian Corn per acre, and would richly repay the husbandman's outlay and efforts. Now, I venture to say that all the grain I saw growing in the valley of the Hudson through Warren County will not return the farmer 75 cents for each day's labor expended thereon, allowing nothing for the use of the land.

"But how shall we obtain fertilizers?" I am often asked. "We are poor; we can afford to keep but few cattle; Guano, Phosphate, Bones, Lime, etc., are beyond our means. Even if we could pay for them, the cost of transportation to our out-of-the-way nooks would be heavy. We cannot deal with our lands so bountifully as you do, but must be content to do as we can."

To all which I make answer: No man ever lacked fertilizers who kept his eyes wide open and devoted two months of each Fall and Winter to collecting and preparing them. Wherever swamp muck may be had, wherever bogs exist or flags or rushes grow, there are materials which, carted into the barn-yard in Autumn or Winter, may be drawn out fertilizers in season for Corn-planting next Spring. Wherever a pond or slough dries up in Summer or Autumn, there is material that may be profitably transformed into next year's grass or grain. In the absence of all these —and they are seldom very far from one who knows how to look for them—rank weeds of all sorts, if cut while green and tender, or forest leaves, gathered in the Fall, used for litter in the stable, and thence thrown into the yard, will serve an excellent purpose. Nay, more: I am confident that the farmer who lacks these, but has access to a bed or bank of simple clay, may cart 200 loads of it in November into an ordinary farm-yard, have it trampled into and mixed with his manure in the Winter, and draw it out in the Spring, excellently fitted to enrich his sandy or gravelly land, and insure him, in connection with deep and thorough culture, a generous yield of Corn, even in such a season as the present. Dr. George B. Loring, the most successful farmer in Massachusetts, uses naked beach sand in abundance as litter for his 80 cows, mixes it with his manure throughout the Winter, and draws out the compound to fertilize his clay meadows in the Spring, with most satisfactory results. Depend on it, no man need lack fertilizers who begins in season and is willing to work for them.

And yet once more :

From the hills which inclose this valley of the upper Hudson (and from ever so many other valleys as well), brooks and rivulets, copious in Spring, when their waters are surcharged and discolored by the richest juices of the uplands, pour down in frequent cascades and dance across the intervale to be lost in the river.   There is scarcely an acre of that intervale which might not be irrigated from these streams at a very moderate outlay of work at the season when work is least pressing : the water thus held back by dams being allowed to flow thence gently and equably across the intervale, conveying not moisture only, but fertility also, to every plant growing thereon.   I am confident that I passed many places on the upper Hudson, as well as on the Connecticut and Ammonoosuc, where 100 faithful days' work providing for irrigation would have given 100 bushels of grain, or 10 tuns of hay additional this year, and as much per annum henceforth, at a cost of not more than two days' work in each year hereafter.

Farmers, but above all farmers' sons, think of these things.

XXXIII.

IF a man whose capital consists of the clothes on his back, $5 in his pocket, and an ax over his right shoulder, undertakes to hew for himself a farm out of the primitive forest, he must of course devote some years to rugged manual labor, or he will fail of success. It is indeed possible that he should find others, even on the rude outposts of civilization, who will hire them to teach school, or serve as county clerk, or survey lands, or do something else of like nature: thus enabling him to do his chopping trees, and rolling logs, and breaking up his stumpy acres, by proxy; but the fair presumption is that he will have to chop and log, and burn off and fence, and break up, by the use of his own proper muscle; and he must be energetic and frugal, as well as fortunate, if he gets a comfortable house over his head, with forty arable acres about him, at the end of fifteen years' hard work. If he has brains, and has been well educated, he may possibly shorten this ordeal to ten years; but, should he begin by fancying hard work beneath him, or his abilities too great to be squandered in bushwhacking, he is very likely to come out at the little end of the horn, and, straggling back to some populous settlement, more needy and seedy than when he set forth to wrest a farm from the wilderness, declare the pioneer's life one of such dreary, hopeless privation that no one who can read or cypher ought ever to attempt it.

A poor man, who undertakes to live by his wits on a farm that he has bought on credit, is not likely to achieve a brilliant success; but the farmer whose hand and brain work in concert will never find nor fancy his intellect or his education too good for his calling. He may very often discover that he wasted months of his school-days on what was ill-adapted to his needs, and of little use in fighting the actual battle of life; but he will at the same time have ample reason to lament the meagerness and the deficiency of his knowledge.

I hold our average Common Schools defective, in that they fail to teach Geology and Chemistry, which in my view are the natural bases of a sound, practical knowledge of things—knowledge which the farmer, of all men, can least afford to miss. However it may be with others, he vitally needs to understand the character and constitution of the soil he must cultivate, the elements of which it is composed, and the laws which govern their relations to each other. Instruct him in the higher mathematics if you will, in logic, in meteorology, in ever so many languages; but not till he shall have been thoroughly grounded in the sciences which unlock for him the arcana of Nature; for these are intimately related to all he must do, and devise, and direct, throughout the whole course of his active career. Whatever he may learn or dispense with, a knowledge of these sciences is among the most urgent of his life-long needs.

Hence, I would suggest that a simple, lucid, lively, accurate digest of the leading principles and facts in Geology and Chemistry, and their application to the practical management of a farm, ought to constitute the Reader of the highest class in every Common School, especially in rural districts. Leave out details and recipes, with directions when to plant or sow, etc.; for these must vary with climates, circumstances, and the progress of knowledge; but let the body and bones, so to speak, of a primary agricultural education be taught in every school, in such terms and with such clearness as to commend them to the understanding of every pupil. I never yet visited a school in which something was not taught which might be omitted or postponed in favor of this.

Out of school and after school, let the young farmer delight in the literature illustrative of his calling—I mean the very best of it. Let him have few agricultural books; but let these treat of principles and laws rather than of methods and applications. Let him learn from these how to ascertain by experiment what are the actual and pressing needs of his soil, and he will readily determine by reflection and inquiry how those needs may be most readily and cheaply satisfied.

All the books in the world never of themselves made one good farmer; but, on the other hand, no man in this age can be a thoroughly good farmer without the knowledge which is more easily and rapidly acquired from books than otherwise. Books are no substitute for open-eyed observation and practical experience; but they enable one familiar with their contents to observe with an accuracy, and experiment with an intelligence, that are unattainable without them. The very farmer who tells you that he never opened a book which treats of Agriculture, and never wants to see one, will ask his neighbor how to grow or cure tobacco, or hops, or sorgho, or any crop with which he is yet unacquainted, when the chances are a hundred to one that this particular neighbor cannot advise him so well as the volume which embodies the experience of a thousand cultivators of this very plant instead of barely one. A good book treating practically of Agriculture, or of some department therein, is simply a compendium of the experience of past ages combined with such knowledge as the present generation have been enabled to add thereto. It may be faulty or defective on some points; it is not to be blindly confided in, nor slavishly followed—it is to be mastered, discussed, criticised, and followed so far as its teachings coincide with the dictates of science, experience, and common sense. Its true office is suggestion; the good farmer will lean upon and trust it as an oracle only where his own proper knowledge proves entirely deficient.

By-and-by, it will be generally realized that few men live or have lived who cannot find scope and profitable employment for all their intellect on a two-hundred-acre farm. And then the farmer will select the brightest of his sons to follow him in the management and cultivation of the paternal acres, leaving those of inferior ability to seek fortune in pursuits for which a limited and special capacity will serve, if not suffice. And then we shall have an Agriculture worthy of our country and the age.

Meantime, let us make the most of what we have, by diffusing, studying, discussing, criticizing, Liebig's Agricultural Chemistry, Dana's Muck Manual, Waring's Elements, and the books that each treat more especially of some department of the farmer's art, and so making ourselves familiar, first, with the principles, then with the methods, of scientific, efficient, successful husbandry. Let us, who love it, treat Agriculture as the elevated, ennobling pursuit it might and should be, and thus exalt it in the estimation of the entire community.

We may, at all events, be sure of this: Just so fast and so far as farming is rendered an intellectual pursuit, it will attract and retain the strongest minds, the best abilities, of the human race. It has been widely shunned and escaped from, mainly because it has seemed a calling in which only inferior capacities were required or would be rewarded. Let this error give place to the truth, and Agriculture will win votaries from among the brightest intellects of the race.

XXXIV.

SHEEP AND WOOL-GROWING.

OURS is eminently an agricultural country. We produce most of our Food, and export much more than we import of both Grain and Meat. Of Cotton, we grow some Three Millions of bales annually, whereof we export fully two-thirds. But of this we reïmport a portion in the shape of Fabrics and of Thread; and yet, while we are largely clothed in Woolens, and extensive sections of our country are admirably adapted to the rearing of Sheep and the production of Wool, we not only import a considerable share of the Woolens in which we are clad, but we also import a considerable proportion of the Wool wherefrom we manufacture the Woolens fabricated on our own soil. In other words: while we are a nation of farmers and herdsmen, we fail to grow so much Wool as is needed to shield us against the caprices and inclemencies of our diverse but generally fitful climates.

There is a seeming excuse for this in the fact that extensive regions in South America and Australia are devoted to Sheep-growing where animals are neither housed nor herded, and where they are exclusively fed, at all seasons, on those native grasses which are the spontaneous products of the soil. I presume Wool is in those regions produced cheaper than it can permanently be on any considerable area of our own soil; and yet I believe that the United States should, and profitably might, grow as much Wool as is needed for their own large annual consumption. Here are my reasons

I. When the predominant interest of British Manufactures constrained the entire repeal of the duties on imported Wool, whereby Sheep-growing had previously been protected, the farmers apprehended that they must abandon that department of their industry; but the event proved this calculation a mistake. They grow more Sheep and at better profit to-day than they did when their Wool brought a higher price under the influence of Protective duties, because the largely increased price of their Mutton more than makes up to them their loss by the reduced prices of their Wool. So, while I do not expect that American Wool will ever again command such high prices as it has done at some periods in the past, I am confident that the general appreciation in the prices of Meat, which has occurred within the last ten or fifteen years, and which seems likely to be enduring, will render Sheep-growing more profitable in the future than it has been in the past. At all events, while our farmers are generally obliged to sell their Grain and Meat at prices somewhat below the range of the British markets, it is hardly conceivable that they should not afford to grow Wool, for which they receive higher average prices than the British farmers do, who feed their Sheep on the produce of lands worth from $300 to $500 (gold) per acre.

II. Interest being relatively high in this country, and Capital with most farmers deficient, it is a serious objection to cattle-growing that the farmer must wait three or four years before receiving a return for his outlay. If he begins poor, with but a few cows and a team, he naturally wants to rear and keep all his calves for several years in order to adequately stock his farm, so that little or no income is meantime realized from his herd; whereas a flock of Sheep yields a fleece per head each year, though not even a lamb is sold, while its increase in numbers is far more rapid than that of a herd of cattle.

III. Almost every farmer, at least in the old States, finds some part of his land infested with bushes and briers, which seem to flourish by cutting, if he finds time to cut them, and which the ruggedness of his soil precludes his exterminating by the plow. In every such case, Sheep are his natural allies—his unpaid police—his vigilant and thorough-going assistants. Give them an even start in Spring with the bushes and briers; let their number be sufficient; and they are very sure to come out ahead in the Fall.

IV. Our farmers in the average are too much confined in Summer and Autumn to salt meats, and especially to Pork. However excellent in quality these may be, their exclusive use is neither healthful nor palatable. With a good flock of Sheep, the most secluded farmer may have fresh meat every week in haying and harvest-time if he chooses; and he will find this better for his family, and more satisfactory to his workmen, than a diet wherefrom fresh meat is excluded.

V. Now, I do not insist that every farmer should grow Sheep, for I know that many are so situated that they cannot. In stony regions, where walls are very generally relied on for fences, I am aware that Sheep are with difficulty kept within bounds; and this is a serious objection. In the neighborhood of cities and large villages, where Fresh Meat may be bought from day to day, one valid reason for keeping them has no application; yet I hold that twice as many of our farmers as now have flocks ought to have them, and would thereby increase their profits as well as the comfort of their families.

The most serious obstacle to Sheep husbandry in this country is the abundance and depredations of dogs. Farmers by tens of thousands have sold off, or killed off, their flocks, mainly because they could not otherwise protect themselves against their frequent decimation by prowling curs, which were not worth the powder required to shoot them. It seems to me that a farmer thus despoiled is perfectly justifiable in placing poisoned food where these cut-throats will be apt to find it while making their next raid on his

Sheep. I should have no scruple in so doing, provided I could guard effectually against the poisoning of any other than the culprits.

In a well-settled, thrifty region, where ample barns are provided, I judge that the losses of Sheep by dogs may be reduced to a minimum by proper precautions. Elsewhere than in wild, new frontier settlements, every flock of Sheep should have a place of refuge beneath the hay-floor of a good barn, and be trained to spend every night there, as well as to seek this shelter against every pelting storm. Even if sent some distance to pasture, an unbarred lane should connect such pasture with their fold; and they should be driven home for a few nights, if necessary, until they had acquired the habit of coming home at nightfall; and I am assured that Sheep thus lodged will very rarely be attacked by dogs or wolves.

As yet, our farmers have not generally realized that enhancement of the value of Mutton, whereby their British rivals have profited so largely. Their fathers began to breed Sheep when a fleece sold for much more than a carcase, and when fineness and abundance of Wool were the main consideration. But such is no longer the fact, at least in the Eastern and Middle States. To-day, large and long-wooled Sheep of the Cotswold and similar breeds are grown with far greater profit in this section than the fine-wooled Merino and Saxony, except where choice specimens of the latter can be sold at high prices for removal to Texas and the Far West. The growing of these high-priced animals must necessarily be confined to few hands. The average farmer cannot expect to sell bucks at $1,000, and even at $5,000, as some have been sold, or at least reported. He must calculate that his Sheep are to be sold, when sold at all, at prices ranging from $10 down to $5, if not lower, so that mechanics and merchants may buy and eat them without absolute ruin; and he must realize that 100 pounds of Mutton at 10 cents, with 6 pounds of Wool at 30 cents, amount to more than 60 pounds of Mutton at 8 cents, and 10 pounds of Wool at 60 cents. Farmers who grow Sheep for Mutton in this vicinity, and manage to have lambs of good size for sale in June or July, assure me that their profit on these is greater than on almost anything else their farms will produce; and they say what they know.

The satisfactory experience of this class may be repeated to-day in the neighborhood of any considerable city in the Union. Sheep-growing is no experiment; it is an assured and gratifying success with all who understand and are fitly placed for its prosecution. Wool may never again be so high as we have known it, since the Far West and Texas can grow it very cheaply, while its transportation costs less than five per cent. of its value, where that of Grain would be 75 per cent.; but Mutton is a wholesome and generally acceptable meat, whereof the use and popularity are daily increasing; so that its market value will doubtless be greater in the future than it has been in the past. I would gladly incite the farmers of our country to comprehend this fact, and act so as to profit by it.

But the new region opened to Sheep-growing by the pioneers of Colorado, and other Territories, is destined to play a great part in the satisfaction of our need of Wool. The elevated Plains and Valleys which enfold and embrace the Rocky Mountains are exceedingly favorable to the cheap production of Wool. Their pure, dry, bracing atmosphere; the rarity of their drenching storms; the fact that their soil is seldom or never sodden with water; and the excellence of their short, thin grasses, even in Winter, render them admirably adapted to the wants of the shepherd and his flocks. I do not believe in the wisdom or humanity, while I admit the possibility, of keeping Sheep without cured fodder on the Plains or elsewhere; on the contrary, I would have ample and effective shelter against cold and wet provided for every flock, with Hay, or Grain, or Roots, or somewhat of each of them, for at least two months of each year; but, even thus, I judge that fine Wool can be grown in Colorado or Wyoming far cheaper than in New England or even Minnesota, and of better quality than in Texas or South America. And I am grievously mistaken if Sheep husbandry is not about to be developed on the Plains with a rapidity and success which have no American precedent.

XXXV.

FARMERS, it is urged, sometimes fail; and this is unfortunately true of them, as of all others. Some fail in integrity; others in sobriety; many in capacity; most in diligence; but not a few in method or system. Quite a number fail because they undertake too much at the outset; that is, they run into debt for more land than they have capital to stock or means to fertilize, and are forced into bankruptcy by the interest ever-accruing upon land which they are unable to cultivate. If they should get ahead a little by active exertion throughout the day, the interest would overtake and pass them during the ensuing night.

Few of the unsuccessful realize the extent to which their ill fortune is fairly attributable to their own waste of time. Men not naturally lazy squander hours weekly in the village, or at the railroad station, without a suspicion that they are thus destroying their chances of success in life. To-day is given up to a monkey-show; half of to-morrow is lost in attendance on an auction; part of next day is spent at

a caucus or a jury trial; and so on until one-third of
the year is virtually wasted.

Now, the men who have achieved eminent success,
within my observation, have all been rigid economists
of time.   They managed to transact their business at
the county-seat while serving there as grand or petit
jurors, or detained under subpœna as witnesses; they
never attended an auction unless they really needed
something which was there to be sold, and then they
began their day's work earlier and ended it later in
order to redeem the time which they borrowed for
the sale.   I do not believe that any American farmer
who could count up three hundred full days' work in
every year between his twenty-first and his thirtieth
ever yet failed, except as a result of speculation, or
endorsing, or inordinate running into debt.

I would, therefore, urge every farmer to keep a
rigid account current of the disposal of his time, so
as to be able to see at the year's end exactly how
many days thereof he had given to productive labor;
how many to such abiding improvements as fencing
and draining; and how many to objects which neither
increased his crop nor improved his farm.   I am sure
many would be amazed at the extent of this last
category.

If every youth who expects to live by farming
would buy a cheap pocket-book or wallet which con-
tains a diary wherein a page is allotted to each day
of the year, and would, at the close of that day, or at
least while its incidents were still fresh in his mind,

set down under its proper head whatever incidents were most noteworthy—as, for instance, a soaking rain; a light or heavy shower; a slight or killing frost; a fall of snow; a hurricane; a hail-storm; a gale; a decidedly hot or notably cold temperature; the turning out of cattle to pasture or sheltering them against the severity of Winter; also the planting or sowing of each crop or field, and whether harm was done to it by frost in its infancy or when it approached maturity—he would thus provide himself with annual volumes of fact which would prove instructive and valuable throughout his maturer years.

The good farmer will of course keep accounts with such of his neighbors as he sees fit to deal with; and he ought to charge a lent or credit a borrowed plow, harrow, reaper, log-chain, or other implement, precisely as though it were meal or meat of an equal value. I judge that borrowed implements, if regularly charged at cost, and credited at their actual value when returned, would generally come home sooner and in better condition.

But the farmer, like every one else, should be most careful to keep debt and credit with himself and his farm. If a dollar is spent or lent, his books should show it; and let items and sum total stare him in the face when he strikes a balance at the close of the year. If there has been no leakage either of dimes or of hours, he will seldom be poorer on the 31st of December than he was on the 1st of the preceding January.

Most farmers fail to keep accounts with their several fields and crops; yet what could be more instructive than these? Here are ten acres of Corn, with a yield of 20 to 40 bushels per acre—a like area and like yield of Oats; a smaller or larger of Rye, Buckwheat, or Beans, as the case may be. If the produce is sold, most farmers know how much it brings; but how many know how much it cost? Say the Corn brings 75 cents per bushel, and the Oats 50 cents: was either or both produced at a profit? If so, at what profit? Here is a farmer who has grown from 100 to 300 bushels of Corn per annum for the last 20 years; ought he not to know by this time what Corn costs him in the average, and whether it could or could not with profit give place to something else? Most farmers grow some crops at a profit, others at a loss; ought they not to know, after an experience of five or ten years, what crops have put money into their pockets, and what have made them poorer for the growing?

Of course, there is complication and some degree of uncertainty in all such account-keeping; for every one is aware that some crops take more from the soil than others, and so leave it in a worse condition for those that are to follow, and that some exact large reënforcements of fertilizers, whereof a part only is fairly chargeable to the first ensuing product, while a large share inures to the subsequent harvests. Each must judge for himself how much is to be credited for such improvement, and how much charged against other crops for deterioration. He, for example, whose meadows will cut from two to three tuns per acre of good English Hay may generally sell that Hay for twice if not thrice the immediate cost of its production, and so seem to be realizing a large profit; but, if he gives nothing to the soil in return for the heavy draft thus made upon it, his crop will dwindle year by year, until it will hardly pay for cutting ; and the diminution in value of his meadows will nearly or quite balance the seeming profit accruing from his Hay. But account-keeping in every business involves essentially identical calculations; and the merchant who this year makes no net profit on his goods, but doubles the number of his customers and the extent of his trade, has thriven precisely as has the farmer whose profit on his crops has all been invested in drains permeating his bogs, and in Lime, Plaster, and other fertilizers, applied to and permanently enriching his dryer fields.

" To make each day a critic on the last," was the aspiration of a wise man, if not a great poet. So the farmer who will keep careful and candid accounts with himself, annually correcting his estimates by the light of experience, will soon learn what crops he may reasonably expect to grow at a profit, and to reject such as are likely to involve him in loss ; and he who, having done this, shall blend common sense with industry, will have no reason to complain thereafter that there is no profit in farming, and no chance of achieving wealth by pursuing it.

XXXVI.

STONE ON A FARM.

THIS earth, geologists say, was once an immense expanse of heated vapor, which, gradually cooling at its surface, as it whirled and sped through space, contracted and formed a crust, which we know as Rock or Stone. This crust has since been broken through, and tilted up into ranges of mountains and hills, by the action of internal fires, by the transmutation of solid bodies into more expansive gases; and the fragments torn away from the sharper edges of upheaved masses of granite, quartz, or sandstone, having been frozen into icebergs floating, or soon to be so, have been carried all over the surface of our planet, and dropped upon the greater part, as those icebergs were ultimately resolved, by a milder temperature, into flowing water. When the seas were afterward reduced nearly or quite to their present limits, and the icebergs restricted to the frigid zones and their vicinity, streams had to make their way down the sides of the mountains and hills to the subjacent valleys and plains, sweeping along not merely sand and gravel, but bowlders also, of every size and form, and sometimes great rocks as well, by the force of their impetuous currents. And, as a very large, if not the larger portion of our earth's surface bears testimony to the existence and powerful action through ages, of larger and smaller water-courses, a wide and general diffusion of stones, not in place, but more or less triturated, smoothed, and rounded, by the action of water, was among the inevitable results.

These stones are sometimes a facility, but oftener an impediment, to efficiency in agriculture. When heated by fervid sunshine throughout the day, they retain a portion of that heat through a part of the succeeding night, thereby raising the temperature of the soil, and increasing the deposit of dew on the plants there growing. When generally broken so finely as to offer no impediment to cultivation, they not merely absorb heat by day, to be given off by night, but, by rendering the soil open and porous, secure a much more extensive diffusion of air through it than would otherwise be possible. Thus do slaty soils achieve and maintain a warmth unique in their respective latitudes, so as to ripen grapes further North, and at higher elevations, than would otherwise be possible.

The great Prairies of the West, with a considerable portion of the valleys and plains of the Atlantic slope, expose no rock at their surfaces, and little beneath them, until the soil has been traversed, and the vicinity of the underlying rock in place fairly attained. To farmers inured to the perpetual stone-picking of New-England, and other hilly regions, this is a most welcome change; but when the pioneer comes to look about him for stone to wall his cellar and his well, to underpin his barn, and form the foundations of his dwelling, he realizes that the bowlders he had exulted in leaving behind him were not wholly and absolutely a nuisance; glad as he was to be rid of them forever, he would like now to call some of them back again.

Yet, the Eastern farmer of to-day has fewer uses for stone than his grandfather had. He does not want his farm cut up into two or three-acre patches, by broad-based, unsightly walls, which frost is apt to heave year after year into greater deformity and less efficiency; nor does he care longer to use them in draining, since he must excavate and replace thrice as much earth in making a stone as in making a tile drain; while the former affords shelter and impunity to rats, mice, and other mischievous, predatory animals, whose burrowing therein tends constantly to stimulate its natural tendency to become choked with sand and earth. Of the stone drains, constructed through parts of my farm by foremen whose wills proved stronger than my own, but two remain in partial operation, and I shall rejoice when these shall have filled themselves up and been counted out evermore. Happily, they were sunk so low that the subsoil plow will never disturb them.

Still, my confidence that nothing was made in vain is scarcely shaken by the prevalence and abundance of stone on our Eastern farms. We may not have present use for them all; but our grandsons will be wiser than we, and have uses for them which we hardly suspect. I rëinsist that land which is very stony was mainly created with an eye to timber-growing, and that millions of acres of such ought forthwith to be planted with Hickory, White Oak, Locust, Chestnut, White Pine, and other valuable forest-trees. Every acre of thoroughly dry land, lying near a railroad, in the Eastern or Middle States, may be made to pay a good interest on from $50 up to $100, provided there be soil enough above its rocks to afford a decent foothold for trees; and how little will answer this purpose none can imagine who have not seen the experiment tried. Sow thickly, that you may begin to cut out poles six to ten feet long within three or four years, and *keep* cutting out (but never cutting off) thenceforward, until time shall be no more, and your rocky crests, steep hill-sides and ravines, will take rank with the most productive portions of your farm.

In the edges of these woods, you may deposit the surplus stones of the adjacent cultivated fields, in full assurance that moth and rust will not corrupt nor thieves break through and steal, but that you and your sons and grandsons will find them there when-ever they shall be needed, as well as those you found there when you came into possession of the farm.

I am further confident that we shall build more and more with rough, unshapen stone, as we grow older and wiser. In our harsh, capricious climate, walls of stone-concrete afford the cheapest and best protection alike against heat and frost, for our animals certainly, and, I think, also for ourselves. Let the farmer begin his barn by making of stone, laid in thin mortar, a substantial basement story, let into a hillside, for his manure and his root-cellar; let him build upon this a second story of like materials for the stalls of his cattle; and now he may add a third story and roof of wood for his hay and grain, if he sees fit. His son or grandson will, probably, take this off, and replace it with concrete walls and a slate roof; or this may be postponed until the original wooden structure has rotted off; but I feel sure that, ultimately, the dwellings as well as barns of thrifty farmers, in stony districts, will mainly be built of rough stone, thrown into a box and firmly cemented by a thin mortar composed of much sand and little lime; and that thus at least ten thousand tuns of stone to each farm will be disposed of. It may be somewhat later still before our barn-yards, fowl inclosures, gardens, pig-pens, etc., will be shut in by cemented walls; but the other sort affords such ample and perpetual lurking-places for rats, minks, weasels, and all manner of destructive vermin, that they are certain to go out of fashion before the close of the next century.

As to blasting out Stone, too large or too firmly fixed to be otherwise handled, I would solve the problem by asking, "Do you mean to keep this lot in cultivation?" If you do, clear it of stone from the surface upward, and for at least two feet downward, though they be as large as haycocks, and as fixed as the everlasting hills. Clear your field of every stone bigger than a goose-egg, that the Plow or the Mower may strike in doing its work, or give it up to timber, plant it thoroughly, and leave its stones unmolested until you or your descendants shall have a paying use for them.

A friend deeply engaged in lumbering gives me a hint, which I think some owners of stony farms will find useful. He is obliged to run his logs down shallow, stony creeks, from the bottom of which large rocks often protrude, arresting the downward progress of his lumber. When the beds of these creeks are nearly dry in Summer, he goes in, with two or three stout, strong assistants, armed with crowbars and levers, and rolls the stones to this side and that, so as to leave a clear passage for his logs. Occasionally, he is confronted by a big fellow, which defies his utmost force ; when, instead of drilling and blasting, he gathers dead tree-tops, and other dry wood of no value, from the banks, and builds a hot fire on the top of each giant bowlder. When the fire has burned out, and the rock has cooled, he finds it softened, and, as it were, rotten, on the top, often split, and every way so demoralized that he can deal with it as though it were chalk or cheese. He estimates his saving by this process, as compared with drilling and blasting, as much more than fifty per cent. I

10

trust farmers with whom wood is abundant, and big stones superabundant, will give this simple device a trial.   Powder and drilling cost money, part of which may be saved by this expedient.

I have built some stone walls—at first, not very well; but for the last ten years my rule has been: Very little fence on a farm, but that little of a kind that asks no forbearance of the wildest bull that ever wore a horn.   The last wall I built cost me at least $5 per rod; and it is worth the money.   Beginning by plowing its bed and turning the two furrows together, so as to raise the ground a foot, and make a shallow ditch on either side, I built a wall thereon which will outlast my younger child.   An ordinary wall dividing a wood on the north from an open field of sunny, gravelly loam on the south, would have been partly thrown down and wholly twisted out of shape in a few years, by the thawing of the earth under its sunny side, while it remained firm as a rock on the north; but the ground is always dry under my entire wall; so nothing freezes there, and there is consequently nothing to thaw and let down my wall.   I shall be sorely disappointed if that wall does not outlast my memory, and be known as a thorough barrier to roving cattle long after the name of its original owner shall have been forgotten.

XXXVII.

FENCES AND FENCING.

THOUGH I have already indicated, incidentally, my decided objections to our prevalent system of Fencing, I deem the subject of such importance that I choose to discuss it directly. Excessive Fencing is peculiarly an American abuse, which urgently cries for reform.

Solon Robinson says the fence-tax is the heaviest of our farmer's taxes. I add, that it is the most needless and indefensible.

Highways we must have, and people must traverse them; but this gives them no right to trample down or otherwise injure the crops growing on either side. In France, and other parts of Europe, you see grass and grain growing luxuriantly up to the very edge of the beaten tracks, with nothing like a fence between them. Yet those crops are nowise injured or disturbed by wayfarers. Whoever chooses to impel animals along these roads must take care to have them completely under subjection, and must see that they do no harm to whatever grows by the way-side.

In this country, cattle-driving, except on a small scale, and for short distances, has nearly been super-

seded by railroads.   The great droves formerly reaching the Atlantic seaboard on foot, from Ohio or further
West, are now huddled into cars and hurried through
in far less time, and with less waste of flesh ; but they
reach us fevered, bruised, and every way unwhole-
some.   Every animal should be turned out to grass,
after a railroad journey of more than twelve hours,
and left there a full month before he is taken to the
slaughter-pen.   We must have many more deaths per
annum in this city than if the animals on which we
subsist were killed in a condition which rendered them
fit for human food.

Ultimately, our fresh Beef, Mutton and Pork, will
come to us from the Prairies in refrigerating cars:
each animal having been killed while in perfect health,
unfevered and untortured by days of cramped, galled,
and thirsty suffering, on the cars.   This will leave
their offal, including a large portion of their bones, to
enrich the fields whence their sustenance was drawn
and from which they should never be taken.   The cost
of transporting the meat, hides, and tallow, in such
cars, would be less than that of bringing through the
animals on their legs ; while the danger of putrefac-
tion might be utterly precluded.

But to return to Fencing:

Our growing plants must be preserved from animal ravage ; but it is most unjust to impose the cost
of this protection on the growers.   Whoever chooses
to rear or buy animals must take care that they do
not infest and despoil his neighbors.   Whoever sees

fit to turn animals into the street, should send some
one with them who will be sure to keep them out of
mischief, which browsing young trees in a forest
clearly is.

If the inhabitants of a settlement or village surrounded by open prairie, see fit to pasture their cattle
thereon, they should send them out each morning in
the charge of a well-mounted herdsmen, whose duty
should be summed up in keeping them from evil-
doing by day and bringing them safely back to their
yard or yards at nightfall.

Fencing bears with special severity on the pioneer
class, who are least able to afford the outlay. The
"clearing" of the pioneer's first year in the wilderness,
being enlarged by ax and fire, needs a new and far
longer environment next year; and so through subsequent years until clearing is at an end. Many a
pioneer is thus impelled to devote a large share of
his time to Fencing; and yet his crops often come to
grief through the depredations of his own or his
neighbor's breachy cattle.

Fences produce nothing but unwelcome bushes,
briers and weeds. So far as they may be necessary,
they are a deplorable necessity. When constructed
where they are not really needed, they evince costly
folly. I think I could point out farms which would
not sell to-day for the cost of rebuilding their present
fences.

We cannot make open drains or ditches serve for
fences in this country, as they sometimes do in milder

and more equable climates, because our severe frosts would heave and crumble their banks if nearly perpendicular, sloping them at length in places so that animals might cross them at leisure. Nor have we, so far north as this city, had much success with hedges, for a like reason. There is scarcely a hedge-plant at once efficient in stopping animals and so hardy as to defy the severity, or rather the caprice, of our Winters. I scarcely know a hedge which is not either inefficient or too costly for the average farmer; and then a hedge is a fixture; whereas we often need to move or demolish our fences.

Wire Fences are least obnoxious to this objection; they are very easily removed; but a careless teamster, a stupid animal, or a clumsy friend, easily makes a breach in one, which is not so easily repaired. Of the few Wire Fences within my knowledge, hardly one has remained entire and efficient after standing two or three years.

Stone Walls, well built, on raised foundations of dry earth, are enduring and quite effective, but very costly. My best have cost me at least $5 per rod, though the raw material was abundant and accessible. I doubt that any good wall is built, with labor at present prices, for less than $3 per rod. Perhaps I should account this costliness a merit, since it must impel farmers to study how to make few fences serve their turn.

Rail Fences will be constructed only where timber is very abundant, of little value, and easily split,

Whenever the burning of timber to be rid of it has ceased, there the making of rail fences must be near its end.

Where fences must still be maintained, I apprehend that posts and boards are the cheapest material. Though Pine lumber grows dear, Hemlock still abounds; and the rapid destruction of trees for their bark to be used in tanning must give us cheap hemlock boards throughout many ensuing years. Spruce, Tamarack, and other evergreens from our Northern swamps, will come into play after Hemlock shall have been exhausted.

As for posts, Red Cedar is a general favorite; and this tree seems to be rapidly multiplying hereabout. I judge that farmers who have it not, might wisely order it from a nursery and give it an experimental trial. It is hardy; it is clean; it makes but little shade; and it seems to fear no insect whatever. It flourishes on rocky, thin soils; and a grove of it is pleasant to the sight—at least, to mine.

Locust is more widely known and esteemed; but the borer has proved destructive to it on very many farms, though not on mine. I like it well, and mean to multiply it extensively by drilling the seed in rich garden soil and transplanting to rocky woodland when two years old. Sowing the seed among rocks and bushes I have tried rather extensively, with poor success. If it germinates at all, the young tree is so tiny and feeble that bushes, weeds, and grass, overtop and smother it.

That a post set top-end down will last many years longer than if set as it grew, I do firmly believe, though I cannot attest it from personal observation. I understand the reason to be this : Trees absorb or suck up moisture from the earth; and the particles which compose them are so combined and adjusted as to facilitate this operation. Plant a post deeply and firmly in the ground, but-end downward, and it will continue to absorb moisture from the earth as it did when alive; and the post, thus moistened to-day and dried by wind and sun to-morrow, is thereby subjected to more rapid disintegration and decay than when reversed.

My general conclusion is, that the good farmer will have fewer and better Fences than his thriftless neighbor, and that he will study and plan to make fewer and fewer rods of fence serve his needs, taking care that all he retains shall be perfect and conclusive. Breechy cattle are a sad affliction alike to their owner and his neighbor; and shaky, rotting, tumble-down fences, are justly responsible for their perverse education. Let us each resolve to take good care that his own cattle shall in no case afflict his neighbors, and we shall all need fewer fences henceforth and evermore.

XXXVIII.

AGRICULTURAL EXHIBITIONS.

I must have attended not less than fifty State or County Fairs for the exhibition (mainly) of Agricultural Machines and Products. From all these, I *should* have learned something, and presume I did; but I cannot now say what. Hence, I conclude that these Fairs are not what they might and should be. In other words, they should be improved. But how?

As the people compose much the largest and best part of these shows, the reform must begin with them. Two-thirds of them go to a Fair with no desire to learn therefrom—no belief that they can there be taught anything. Of course, not seeking, they do not find. If they could but realize that a Farmer's Fair might and should teach farmers somewhat that would serve them in their vocation, a great point would be gained. But they go in quest of entertainment, and find this mainly in horse-racing.

. Of all human opportunities for instruction in humility and self-depreciation, the average public speaker's is the best. He hurries to a place where he has been told that his presence and utterance are earnestly and

generally desired—perhaps to find that his invitation
came from an insignificant and odious handful, who
had some private ax to grind so repugnant to the
great majority that they refuse to countenance the
procedure, no matter how great the temptation.
Even where there is no such feud, many, having
satiated their curiosity by a long stare at him, walk
whistling off, without waiting or wishing to hear him.
But the speaker at a Fair must compete with a thousand counter-attractions, the least of them far more
popular and winning than *he* can hope to be. He is
heard, so far as he is heard at all, in presence of and
competition with all the bellowing bulls, braying
jacks, and squealing stallions, in the county; if he
holds, nevertheless, a quarter of the crowd, he does
well: but let two jockeys start a buggy-race around
the convenient track, and the last auditor shuts his
ears and runs off to enjoy the spectacle. Decidedly,
I insist that a Fair-ground is poorly adapted to the
diffusion of Agricultural knowledge—that the people
present acquire very little information there, even
when they get all they want.

What is needed to render our annual Fairs useful
and instructive far beyond precedent, I sum up as
follows:

I. Each farmer in the county or township should
hold himself bound to make *some* contribution thereto. If only a good hill of Corn, a peck of Potatoes,
a bunch of Grapes, a Squash, a Melon, let him send
that. If he can send all of these, so much the better.

There is very rarely a thrifty farmer who could not add to the attractions and merits of a Fair if he would try. If he could send a coop of superior Fowls, a likely Calf, or a first-rate Cow, better yet; but nine-tenths of our farmers regard a Fair as something wherewith they have nothing to do, except as spectators. When it is half over, they lounge into it with hands in their pockets, stare about for an hour, and go home protesting that they could beat nearly everything they saw there. Then why did they not try? How can we have good Fairs, if those who might make the best display of products save themselves the trouble by not making any? The average meagerness of our Fairs, so generally and justly complained of, is not the fault of those who sent what they had, but of those who, having better, were too lazy to send anything. Until this is radically changed, and the blame fastened on those who might have contributed, but did not, our Fairs cannot help being generally meager and poor.

II. It seems to me that there is great need of an interesting and faithful running commentary on the various articles exhibited. A competent person should be employed to give an hour's off-hand talk on the cattle and horses on hand, explaining the diverse merits and faults of the several breeds there exhibited, and of the representatives of those breeds then present. If any are peculiarly adapted to the locality, let that fact be duly set forth, with the simple object of enabling the farmers to breed more intelligently, and more profitably. Then let the implements and machinery on exhibition be likewise explained and discussed, and let their superiority in whatever respect to those they have superseded or are designed to supersede be clearly pointed out. So, if there be any new Grain, Vegetable, or Fruit, on the tables, let it be made the subject of capable and thoroughly impartial discussion, before such only as choose to listen, and without putting the mere sight-seers to grave inconvenience. A lecture-room should always be attached to a Fair-ground, yet so secluded as to shut out the noise inseparable from a crowded exhibition. Here, meetings should be held each evening, for general discussion; every one being encouraged to state concisely the impressions made on him, and the improvements suggested to him, by what he had seen. Do let us try to reflect and consider more at these gatherings, even though at the cost of seeing less.

III. The well supported Agricultural Society of a rich and populous county must be able, or should be able, to give two or three liberal premiums for general proficiency in farming. If $100 could be proffered to the owner or manager of the best tilled farm in the county, $50 to the owner of the best orchard, and $50 to the boy under 18 years of age who grew the best acre of Corn or Roots that year, I am confident that an impulse would thereby be given to agricultural progress. Our premiums are too numerous and too petty, because so few are willing to contribute with no expectation of personal benefit or distinction. If we had but the right spirit aroused, we might dispense with most of our petty premiums, or replace them by medals of no great cost, and devote the money thus saved to higher and nobler ends.

IV. Much of the speaking at Fairs seems to me insulting to the intelligence of the Farmers present, who are grossly flattered and eulogized, when they often need to be admonished and incited to mend their ways. What use or sense can there be in a lawyer, doctor, broker, or editor, talking to a crowd of farmers as if they were the most favored of mortals and their life the noblest and happiest known to mankind? Whatever it might be, and may yet become, we all know that the average farmer's life is not what it is thus represented: for, if it were, thousands would be rushing into it where barely hundreds left it : whereas we all see that the fact is quite otherwise. No good can result from such insincere and extravagant praises of a calling which so few freely choose, and so many gladly shun. Grant that the farmer's *ought* to be the most enviable and envied vocation, we know that in fact it *is* not; and, agreeing that it should be, the business in hand is to make it so. There must be obstacles to surmount, mistakes to set right, impediments to overcome, before farming can be in all respects the idolized pursuit which poets are so ready to proclaim it and orators so delight to represent it. Let us struggle to make it all that fancy has ever painted it; but, so long as it is not, let us respect undeniable facts, and characterize it exactly as it is.

V. If our counties were thoroughly canvassed by township committees, and each tiller of the soil asked to pledge himself in writing to exhibit something at the next County Fair, we should soon witness a decided improvement. Many would be incited to attend who now stay away; while the very general complaint that there is nothing worth coming to see would be heard no more. As yet, a majority of farmers regard the Fair much as they do a circus or traveling menagerie, taking no interest in it except as it may afford them entertainment for the passing hour. We must change this essentially; and the first step is to induce, by concerted solicitation, at least half the farmers in the county to pledge themselves each to exhibit something at the next annual Fair, or pay $5 toward increasing its premiums.

VI. In short, we must all realize that the County or Township Fair is *our* Fair—not got up by others to invite our patronage or criticism, but something whereto it is incumbent on us to contribute, and which must be better or worse as we choose to make it. Realizing this, let us stop carping and give a shoulder to the wheel.

XXXIX.

I AM not a scientific farmer; it is not probable that I ever shall be. I have no such knowledge of Chemistry and Geology as any man needs to make him a thoroughly good farmer. I am quite aware that men have raised good crops—a good many of them—who knew nothing of science, and did not consider any acquaintance with it conducive to efficiency or success in their vocation. I have no doubt that men will continue to grow such crops, and to make money by agriculture, who hardly know what is meant by Chemistry or Geology ; and yet I feel sure that, as the years roll by, Science will more and more be recognized and accepted as the true, substantial base of efficient and profitable cultivation. Let me here give briefly the grounds of this conviction :

Every plant is composed of elements whereof a very small portion is drawn from the soil, while the ampler residue, so long as the plant continues green and growing, is mainly water, though a variable and often considerable proportion is imbibed or absorbed from the atmosphere, which is understood to yield freely nearly all the elements required of it, provided the plants are otherwise in healthful and thrifty condition.   Water is supplied from the sky, or from springs and streams ; and little more than the most ordinary capacity for observation is required to determine when it is present in sufficient quantity, when in baleful excess.   But who, unaided by Science, can decide whether the soil does or does not contain the elements requisite for the luxuriant growth and perfect development of Wheat, or Fruit, or Grass, or Beets, or Apples?   Who knows, save as he blindly infers from results, what mineral ingredients of this or that crop are deficient in a given field, and what are present in excess?   And how shall any one be enlightened and assured on the point, unless by the aid of Science?

I have bought and applied to my farm some two thousand bushels of Lime, and ten or a dozen tuns of Plaster ; and I infer, from what seemed to be results, that each of these minerals has been applied with profit; but I do not *know* it.   The increased product which I have attributed to one or both of these elements may have had a very different origin and impulse.   I only grope my way in darkness when I should clearly and surely see.

An agricultural essayist in Maine has recently put forth a canon which, if well grounded, is of great value to farmers.   He asserts that the growth of acid plants like Sorrel, Dock, etc., in a field, results from sourness in the soil, and that, where this exists, Lime

—that is, the ordinary Carbonate of Lime—is urgently required; whereas the application of Plaster or Gypsum (Sulphate of Lime) to that field must be useless and wasteful. If such be the truth, a knowledge of it would be worth millions of dollars to our farmers. But I lack the scientific attainment needed to qualify me for passing judgment thereon.

There is great diversity of opinion among farmers with regard to the value of Swamp Muck. One has applied it to his land to good purpose; so he holds Muck, if convenient, the cheapest and best fertilizer a farmer can add to his ordinary barn-yard manure; another has applied cords upon cords of Muck, and says he has derived therefrom no benefit whatever. Now, this contrariety of conclusion may result from imperfect judgment on one side or the other, or from the condition precedent of the diverse soils : one of them requiring what Muck could supply, while the other required something very different from that; or it may be accounted for by the fact that the Muck applied in one case was of superior quality, and in the other good for nothing. Where Muck is composed almost wholly of the leaves of forest-trees which, through thousands of years, have been blown into a bog, or shallow pond, and there been gradually transformed into a fine, black dust or earth, I do not see how it can possibly be applied to an upland, especially a sandy or gravelly soil, without conducing to the subsequent production of bounteous crops. True, it may be sour when first drawn from the stag-

nant pool or bog in which it has lain so long, and
may need to be mixed with Lime, or Salt, or Ashes,
and subjected to the action of sun and frost, to ripen
and sweeten it.   But it seems to me impossible that
such Muck should be applied to almost any reasonably dry land, without improving its consistency and
increasing its fertility.   But all Muck is not the product of decayed forest-leaves; and that which was
formed of coarse, rank weeds and brakes, of rotten
wood and flags, or skunk cabbage, may be of very inferior quality, so as hardly to repay the cost of digging and applying it.   Science will yet enable us to
fix, at least approximately, the value of each deposit
of Muck, and so give a preference to the best.

The Analysis of Soils, whereof much was heard
and whence much was hoped a few years since, seems
to have fallen into utter discredit, so that every
would-be popular writer gives it a passing fling or
kick.   That any analysis yet made was and is worthless, I can readily concede, without shaking in the least
my conviction that soils will yet be analyzed, under
the guidance of a truer, profounder Science, to the
signal enlightenment and profit of their cultivators.
Here is a retired merchant, banker, doctor, or lawyer,
who has bought a spacious and naturally fertile but
worn-out, run-down farm, on which he proposes to
spend the remainder of his days.   Of course, he must
improve and enrich it; but with what? and how?
All the manure he finds, or, for the present, can make
on it, will hardly put the first acre in high condition,

while he grows old and is unwilling to wait forever. He is able and ready to buy fertilizers, and does buy right and left, without knowing whether his land needs Lime, or Phosphate, or Potash, or something very different from either. Say he purchases $2,000 worth of one or more of these fertilizers : it is highly probable that $1,500 might have served him better if invested in due proportion in just what his land most urgently needs; and I unflinchingly believe that we shall yet have an analysis of soils that will tell him just what fertilizers he ought to apply, and what quantity of each of them.

Science has already taught us that every load of Hay or Grain drawn from a field abstracts therefrom a considerable quantity of certain minerals — say Potash, Lime, Soda, Magnesia, Chlorine, Silica, Phosphorus—and that the soil is thereby impoverished until they be replaced, in some form or other. As no deposit in a bank was ever so large that continual drafts would not ultimately exhaust it, so no soil was ever so rich that taking crop after crop from it annually, yet giving nothing back, would not render it sterile or worthless. Sun and rain and wind will do their part in the work of renovation; but all of them together cannot restore to the soil the mineral elements whereof each crop takes a portion, and which, being once completely exhausted, can only be replaced at a heavy cost. Science teaches us to foresee and prevent such exhaustion—in part, by a rotation of crops, and in part by a constant replacement of the minerals annually borne away: the subtraction being greater in proportion as the crop is more exacting and luxuriant.

What I know of Science as applicable to Farming is little indeed ; but I know that there *is* such Science, and that each succeeding year enlarges, improves, and perfects it. I know that I should thus far have farmed to far better purpose, if I had been master even of so much Science as already exists.

Understand that I am not a teacher of this Science —I stand very low in the class of learners. I began to learn too late in life, and have been too incessantly harassed by a multiplicity of cares, to make any satisfactory progress. Any tolerably educated boy of fifteen may know far more of Agricultural Science by the time he has passed his eighteenth birth-day than I do. What I know in this respect can help him very little ; my faith that there is much to be known, and that he may master it if he will, is all that is of much importance. If I can convince a considerable number of our youth that they may surely acquire a competence by the time they shall have passed their fortieth year, without excessive labor or penurious frugality, by means of that knowledge of principles and laws subservient to Agriculture which their fathers could not, but which they easily may attain, I shall have rendered a substantial service alike to them and to our country.

XL.

FARM IMPLEMENTS.

A GOOD workman, it is said, does not quarrel with his tools—which, if true, I judge is due to the fact that he generally manages to have good ones. To work hard throughout a long day under a burning sun, is sufficiently trying, without rendering the labor doubly repugnant by the use of ill-contrived, imperfect, inefficient implements.

The half-century which nearly bounds my recollection has witnessed great improvements in this respect. The Plow, mainly of wood, wherewith my father broke up his stony, hide-bound acres of New-Hampshire pebbles and gravel, in my early boyhood, would now be spurned if offered as a gift to the poorest and most thriftless farmer among us ; and the Hoes which were allotted to us boys in those days, after the newer and better had been assigned to the men, would be rejected with disdain by the stupidest negro in Virginia. Though there is still room for improvement, we use far better implements than our grandfathers did, with a corresponding increase in the efficiency of our labor; but the cultivators of Spain, Portugal, and the greater part of Europe, still linger in the dark ages in this respect. Their plows are little better than the forked sticks which served their barbarian ancestors, and their implements generally are beneath contempt. With such implements, deep and thorough culture is simply impossible, unless by the use of the spade; and he must be a hard worker who produces a peck of Wheat or half a bushel of Indian Corn per day by the exclusive use of this tool. The soil of France is so cut up and subdivided into little strips of two or three roods up to as many acres each—each strip forming the entire patrimony of a family—that agricultural advancement or efficiency is, with the great mass of French cultivators, out of the question. Hence, I judge that, outside of Great Britain and Australia, there is no country wherein an average year's work produces half so much grain as in our own, in spite of our slovenly tillage, our neglect and waste of fertilizers, and the frequent failures of our harvests. Belgium, Holland, and northern France, can teach us neatness and thoroughness of cultivation; the British isles may fairly boast of larger and surer crops of Wheat, Oats, Potatoes, and Grass, than we are accustomed to secure; but, in the selection of implements, and in the average efficiency of labor, our best farmers are ahead of them all.

Bear with me, then, while I interpose a timid plea for our inventors and patentees of implements, whose solicitations that a trial, or at least an inspection, be accorded to their several contrivances, are too often repelled with churlish rudeness. I realize that our thriving farmers are generally absorbed in their own plans and efforts, and that the agent or salesman who insists on an examination of his new harrow, or pitch-fork, or potato-digger, is often extravagant in his assumptions, and sometimes a bore. Still, when I recollect how tedious and how back-breaking were the methods of mowing Grass and reaping Grain with the Scythe and Sickle, which held unchallenged sway in my early boyhood, I entreat the farmer who is petitioned to accord ten or fifteen minutes to the setting forth, by some errant stranger, of the merits of his new horse-hoe or tedder, to give the time, if he can; and that without sour looks or a mien of stolid incredulity. The Biblical monition that, in evincing a generous hospitality, we may sometimes entertain angels unawares, seems to me in point. A new implement may be defective and worthless, and yet contain the germ or suggest the form of a thoroughly good one. Give the inventor or his representative a courteous hearing if you can, even though this should constrain you to make up the time so lost after the day's work would otherwise have ended.

I suspect that the average farmer of our completely rural districts would be surprised, if not instructed, by a day's careful scrutiny of the contents of one of our great implement warehouses. So many and such various and ingenious devices for pulverizing the earth applying fertilizers to the soil, planting or sowing rapidly, eradicating weeds, economizing labor in harvesting, etc., will probably transcend not merely his experience, but his imagination; and every one of these myriad implements is useful in its place, though no single farmer can afford to buy all or half of them. It will yet, I think, be found necessary by the farmers of a school-district, if not of a township, to meet and agree among themselves that one will buy this implement, another that, and so on, until twenty or thirty such devices as a Stump or Rock-Puller, a Clod-Crusher, Thrashing-Machine, Fanning-Mill, etc., shall be owned in the neighborhood—each by a separate farmer, willing to live and let live—with an understanding that each shall be used in turn by him who needs it; and so every one shall be nearly as well accommodated as though he owned them all.

For the number and variety of useful implements increase so rapidly, while their usefulness is so palpable, that, though it is difficult to farm efficiently without many if not most of them, it is impossible that the young farmer of moderate means should buy and keep them all. True, he might hire when he needed, if what he wanted were always at hand; but this can only be assured by some such arrangement as I have suggested, wherein each undertakes to provide and keep that which he will most need; agreeing to lend it whenever it can be spared to any other member of the combination, who undertakes to minister in like manner to *his* need in return.

I think few will doubt that the inventions in aid of Agriculture during the last forty years will be far surpassed by those of the forty years just before us. The magnificent fortunes which, it is currently understood, have rewarded the inventors of the more popular Mowers, Reapers, etc., of our day, are sure to stimulate alike the ingenuity and the avarice of clever men throughout the coming years, and to call into existence ten thousand patents, whereof a hundred will be valuable, and ten or twelve eminently useful. Plowing land free from stumps and stones cannot long be the tedious, patience-trying process we have known it. The machinery which will at once pulverize the soil to a depth of two feet, fertilize and seed it, not requiring it to be trampled by the hoofs of animals employed in subsoiling and harrowing, will soon be in general use, especially on the spacious, deep, inviting prairies of the Great West.—But I must defer what I have to say of Steam and its uses in Agriculture to another chapter.

XLI.

STEAM IN AGRICULTURE.

As yet, the great body of our farmers have been slow in availing themselves of the natural forces in operation around them. Vainly for them does the wind blow across their fields and over their hill-tops. It neither thrashes nor grinds their grain; it has ceased even to separate it from the chaff. The brook brawls and foams idly adown the precipice or hill-side: the farmer grinds his grain, churns his cream, and turns his grindstone, just as though falling water did not embody power. He draws his Logs to one mill, and his Wheat, Corn, or Rye to another, and returns in due season with his boards or his meal; but the lesson which the mill so plainly teaches remains by him unread. Where running or leaping water is not, there brisk breezes and fiercer gales are apt to be. But the average farmer ignores the mechanical use of stream and breeze alike, taxing his own muscle to achieve that which the blind forces of Nature stand ready to do at his command. It may not, and I think it will not, be always thus.

Steam, as a cheap source of practically limitless power, is hardly a century old; yet it has already revolutionized the mechanical and manufacturing industry of Christendom. It weaves the far greater part of all the Textile Fabrics that clothe and shelter and beautify the human family. It fashions every bar and every rail of Iron or of Steel; it impels the machinery of nearly every manufactory of wares or of implements; and it is very rapidly supplanting wind in the propulsion of vessels on the high seas, as it has already done on rivers and on most inland waters.

Water is, however, still employed as a power in

certain cases, but mainly because its adaptation to this end has cost many thousands of dollars which its disuse would render worthless.

I am quite within bounds in estimating that nine-tenths of all the material force employed by man in Manufactures, Mechanics, and Navigation, is supplied by Steam, and that this disproportion will be increased to ninety-nine hundredths before the close of this century.

For Agriculture, Steam has done very much, in the transportation of crops and of fertilizers, but very little in the preparation or cultivation of the soil. Of steam-wagons for roads or fields, steam-plows for pulverizing and deepening the soil, and steam-cultivators for keeping weeds down and rendering tillage more efficient, we have had many heralded in sanguine bulletins throughout the last forty years, but I am not aware that one of them has fulfilled the sanguine hopes of its author. Though a dozen Steam-Plows have been invented in this country, and several imported from Europe, I doubt that a single square mile of our country's surface has been plowed wholly by steam down to this hour. If it has, Louisiana—a State which one would not naturally expect to find in the van of industrial progress—has enjoyed the benefit and earned the credit of the achievement.

Of what Steam has yet accomplished in direct aid of Agriculture, I have little to say, though in Great Britain quite a number of steam-plows are actually at work in the fields, and (I am assured) with fair success. Until something breaks or gives out, one of these plows does its appointed work better and cheaper than such work is or can be done by animal power ; but all the steam-plows whereof I have any knowledge seem too bulky, too complicated, too costly, ever to win their way into general use. I value them only as hints and incitements toward something better suited to the purpose.

What our farmers need is not a steam-plow as a specialty, but a locomotive that can travel with facility, not only on common wagon-roads, but across even freshly-plowed fields, without embarrassment, and prove as docile to its manager's touch as an average span of horses. Such a locomotive should not cost more then $500, nor weigh more than a tun when laden with fuel and water for a half-hour's steady work. It should be so contrived that it may be hitched in a minute to a plow, a harrow, a wagon, or cart, a saw or grist-mill, a mower or reaper, a thresher or stalk-cutter, a stump or rock-puller, and made useful in pumping and draining operations, digging a cellar or laying up a wall, as also in ditching or trenching. We may have to wait some years yet for a servant so dexterous and docile, yet I feel confident that our children will enjoy and appreciate his handiwork.

The farmer often needs far more power at one season than at another, and is compelled to retain and subsist working animals at high cost through months in which he has no use for them, because he must have them when those months have transpired. If he could replace those animals by a machine which, when its season of usefulness was over, could be cleaned, oiled, and put away under a tight roof until next seeding-time, the saving alike of cost and trouble would be very considerable.

When our American reapers first challenged attention in Great Britain, the general skepticism as to their efficiency was counteracted by the suggestion that, even though reaping by machinery should prove more expensive than reaping by hand, the ability to cut and save the grain-crop more rapidly than hitherto would overbalance that enhancement of cost. In the British Isles, day after day of chilling wind and rain is often encountered in harvest-time : the standing Wheat or Oats or Barley becoming draggled, or lodged, or beaten out, while the owner impatiently awaits the recurrence of sunny days. When these at length arrive, he is anxious to harvest many acres at once, since his Grain is wasting and he knows not how soon cloud and tempest may again be his portion. But all his neighbors are in like predicament with himself, and all equally intent on hurrying the harvest; so that little extra help is attainable. If now the aid of a machine may be commanded, which will cut 15 or 20 acres per day, he cares less how much that work will cost than how soon it can be effected. Hence, even though cutting by horse-power had proved more costly than cutting by hand, it would still have been welcome.

So it is with Plowing, here and almost everywhere.
Our farmers have this year been unable to begin
Plowing for Winter Grain so early as they desired,
by reason of the intense heat and drouth, whereby
their fields were baked to the consistency of half-
burned brick.   Much seed will in consequence have
been sown too late, while much seeding will have
been precluded altogether, by inability to prepare the
ground in due season.   If a machine had been at
hand whereby 15 or 20 acres per day could have been
plowed and harrowed, thousands would have invoked
its aid to enable them to sow their Grain in tolerable
season, even though the cost had been essentially
heavier than that of old-fashioned plowing.   I traversed Illinois on the 13th and 14th of May, 1859,
when its entire soil seemed soaked and sodden with
incessant rains, which had not yet ceased pouring.
Inevitably, there had been little or no plowing yet
for the vast Corn-crop of that State ; yet barely two
weeks would intervene before the close of the proper
season for Corn-planting.   Even if these should be
wholly favorable, the plowing could not be effected
in season, and much ground must be planted too late
or not planted at all.   In every such case, a machine
that would plow six or eight furrows as fast as a man
ought to walk, would add immensely to the year's
harvest, and be hailed as a general blessing.

I recollect that a German observer of Western cultivation—a man of decided perspicacity and wide
observation—recommended that each farmer who had

not the requisite time or team for getting in his Corn-crop in due season should plow single furrows through his field at intervals of 3 to $3\frac{1}{2}$ feet, plant his Corn on the earth thus turned, and proceed, so soon as his planting was finished, to plow out the spaces as yet undisturbed between the springing rows of Corn. I do not know that this recommendation was ever widely followed; but I judge that, under certain circumstances, it might be, to decided advantage and profit.

I have not attempted to indicate all the benefits which Steam is to confer directly on Agriculture, within the next half-century. That Irrigation must become general, I confidently believe; and I anticipate a very extensive sinking of wells, at favorable points, in order that water shall be drawn therefrom by wind or steam to moisten and enrich the slopes and plains around them. Such a locomotive as I have foreshadowed might be taken from well to well, pumping from each in an hour or two sufficient water to irrigate several of the adjacent acres; thus starting a second crop of Hay on fields whence the first had been taken, and renewing verdure and growth where we now see vegetation suspended for weeks, if not months. I feel sure that the mass of our farmers have not yet realized the importance and beneficence of Irrigation, nor the facility wherewith its advantages may be secured.

XLII

CO-OPERATION IN FARMING.

THE word of hope and cheer for Labor in our days is COÖPERATION—that is, the combination by many of their means and efforts to achieve results beneficial to them all. It differs radically from Communism, which proposes that each should receive from the aggregate product of human labor enough to satisfy his wants, or at least his needs, whether he shall have contributed to that aggregate much, or little, or nothing at all. Coöperation insists that each shall receive from the joint product in proportion to his contributions thereto, whether in capital, skill, or labor. If one associate has ten children and another none, Communism would apportion to each according to the size of his family alone; while Coöperation would give to each what he had earned, regardless of the number dependent upon him. Thus the two systems are radical antagonists, and only the grossly ignorant or willfully blind will confound them.

A young farmer, whose total estate is less than $500, not counting a priceless wife and child, resolves to migrate from one of the old States to Kansas, Minnesota, or one of the Territories: he has heard that he will there find public land whereon he may make a home of a quarter-section, paying therefor $20 or less for the cost of survey and of the necessary papers. So he may: but, on reaching the Land of Promise, whether with or without his family, he finds a very large belt of still vacant land beyond the settlements already transformed into private property, and either not for sale at all or held on speculation, quite out of his reach. The public land which he may take under the Homestead law lies a full day's journey beyond the border settlements, to which he must look for Mills, Stores, Schools, and even Highways. If he persists in squatting, with intent to earn his quarter-section by settlement and cultivation, he must take a long day's journey across unbridged streams and sloughs, over unmade roads, to find boards, or brick, or meal, or glass, or groceries; while he must postpone the education of his children to an indefinite future day. Gradually, the region will be settled, and the conveniences of civilization will find their way to his door, but not till after he will have suffered through several years for want of them; often compelled to make a journey to get a plow or yoke mended, a grist of grain ground, or to minister to some other trivial but inexorable want. He who thus acquires his quarter-section must fairly earn it, and may be thankful if his children do not grow up rude, coarse, and illiterate.

11*

But suppose one thousand just such young farmers as he is, with no more means and no greater efficiency than his, were to set forth together, resolved to find a suitable location whereon they might all settle on adjoining quarter-sections, thus appropriating the soil of five or six embryo townships : who can fail to see that three-fourths of the obstacles and discouragements which confront the solitary pioneer would vanish at the outset ? Roads, Bridges, Mills,—nay, even Schools and Churches—would be theirs almost immediately ; while mechanics, merchants, doctors, etc., would fairly overrun their settlement and solicit their patronage at every road-crossing. Within a year after the location of their several claims, they would have achieved more progress and more comfort than in five years under the system of straggling and isolated settlement which has hitherto prevailed. The change I here indicate appeals to the common sense and daily experience of our whole people. It is not necessary, however desirable, that the pioneers should be giants in wisdom, in integrity, or in piety, to secure its benefits. A knave or a fool may be deemed an undesirable neighbor; but a dozen such in the township would not preclude, and could hardly diminish, the advantages naturally resulting from settlement by Coöperation.

Nor are these confined to pioneers transcending the boundaries of civilization. I wish I could induce a thousand of our colored men now precariously subsisting by servile labor in the cities, to strike out boldly for homes of their own, and for liberty to direct their own labor, whether they should settle on the frontier in the manner just outlined, or should buy a tract of cheap land on Long Island, in New-Jersey, Maryland, or some State further South. I cannot doubt that the majority of them would work their way up to independence; and this very much sooner, and after undergoing far less privation, than almost every pioneer who has plunged alone into the primitive forest or struck out upon the broad prairie and there made himself a farm.

The insatiable demand for fencing is one of the pioneer's many trials. Though he has cleared off but three acres of forest during his first Fall and Winter, he must surround those acres with a stout fence, or all he grows will be devoured by hungry cattle—his own, if no others. Whether he adds two or ten acres to his clearing during the next year, they must in turn be surrounded by a fence; and nothing short of a very stout one will answer: so he goes on clearing and fencing, usually burning up a part of his fence whenever he burns over his new clearing; then building a new one around this, which will have to be sacrificed in its turn. I believe that many pioneers have devoted as much time to fencing their fields as to tilling them throughout their first six or eight years.

It is different with those who settle on broad prairies, but not essentially better. Each pioneer must fence his patch of tillage with material which

costs him more, and is procured with greater diffi
culty, than though he were cutting a hole in the
forest. Often, when he thinks he has fenced sufficiently, the hungry, breachy cattle, who roam the
open prairies around him, judge his handiwork less
favorably; and he wakes some August morning,
when feed is poorest outside and most luxuriant
within his inclosure, to find that twenty or thirty
cattle have broken through his defenses and half destroyed his growing crop.

If, instead of this wasteful lack of system, a thousand or even a hundred farmers would combine to
fence several square miles into one grand inclosure
for cultivation, erecting their several habitations
within or without its limits, as to each should be convenient—apportioning it for cultivation, or owning
it in severalty, as they should see fit—an immense
economy would be secured, just when, because of
their poverty, saving is most important. Their stock
might range the open prairie unwatched; and they
might all sleep at night in serene confidence that their
corn and cabbages were not in danger of ruthless destruction. Among the settlers in our great primitive
forests, the system of Coöperative Farming would
have to be modified in details, while it would be in
essence the same.

And, once adopted with regard to fencing, other
adaptations as obvious and beneficent would from
day to day suggest themselves. Each pioneer would
learn how to advance his own prosperity by com-

bining his efforts with those of his neighbors. He would perceive that the common wants of a hundred may be supplied by a combined effort at less than half the cost of satisfying them when each is provided for alone. He would grow year by year into a clearer and firmer conviction that short-sighted selfishness is the germ of half the evils that afflict the human race, and that the true and sure way to a bounteous satisfaction of the wants of each is a generous and thoughtful consideration for the needs of all.

And here let me pay my earnest and thankful tribute to Mr. E. V. de Boissière, a philanthropic Frenchman, who has purchased 3,300 acres of mainly rolling prairie-land in Kansas, near Princeton, Franklin County, and is carefully, cautiously, laying thereon the foundations of a great coöperative farm, where, in addition to the usual crops, it is expected that Silk and other exotics will in due time be extensively grown and transformed into fabrics, and that various manufactures will vie with Agriculture in affording attractive and profitable employment to a considerable population. I have not been accustomed to look with favor on our new States and unpeopled Territories as an arena for such experiments, since so many of their early settlers are intent on getting rich by land-speculation—at all events, through the exercise of some others' muscles than their own—while the oppor-

tunities for and incitements to migration and re-
location are so multiform and powerful. Doubtless, M. de Boissière will be often tried by stam-
pedes of his volunteer associates, who, after the
novelty of coöperative effort has worn off, will
find life on his domain too tame and humdrum for
their excitable and high-strung natures. I trust,
however, that he will persevere through every discouragement, and triumph over every obstacle; that
the right men for associates will gradually gather
about him; that his enterprise and devotion will
at length be crowned by a signal and inspiring success; and that thousands will be awakened by it to a
larger and nobler conception of the mission of Industry, and the possibilities of achievement which
stud the path of simple, honest, faithful, persistent
Work.

XLIII.

FARMERS' CLUBS.

FARMERS like other men, divide naturally into two
classes—those who do too much work, and those who
do too little. I know men who are no farmers at all,
only by virtue of the fact that each of them inherited,
or somehow acquired, a farm, and have since lived
upon and out of it, in good part upon that which it
could not help producing—they not doing so much as

one hundred fair days' work each per annum. One of this class never takes a periodical devoted to farming; evinces no interest in county fairs or township clubs, save as they may afford him an excuse for greater idleness; and insists that there is no profit in farming. As land steadily depreciates in quality under his management, he is apt to sell out whenever the increase of population or progress of improvement has given additional value to his farm, and move off in quest of that undiscovered country where idleness is compatible with thrift, profits are realized from light crops, and men grow rich by doing nothing.

The opposite class of wanderers from the golden mean is hardly so numerous as the idlers, yet it is quite a large one. Its leading embodiment, to my mind, is one whom I knew from childhood, who, born poor and nowise favored by fortune, was rated as a tireless worker from early boyhood, and who achieved an independence before he was forty years old in a rural New-England township, simply by rugged, persistent labor—in youth on the farms of other men; in manhood, on one of his own. This man was older at forty than his father, then seventy, and died at fifty, worn out with excessive and unintermitted labor, leaving a widow who greatly preferred him to all his ample wealth, and an only son who, so soon as he can get hold of it, will squander the property much faster, and even more unwisely, than his father acquired it.

To the class of which this man was a fair representative, Farmers' Clubs must prove of signal value.
Though there should be nothing else than a Farmers'
Club in his neighborhood, it can hardly fail in time
to make such a one realize that life need not and
should not be all drudgery ; that there are other
things worth living for beside accumulating wealth.
Let his wife and his neighbor succeed in drawing
such a one into two or three successive meetings, and
he can hardly fail to perceive that thrift is a product
of brain as well as of muscle ; that he may grow rich
by learning and knowing as well as by delving, and
that, even though he should not, there are many
things desirable and laudable beside the accumulation
of wealth.

A true Farmers' Club should consist of all the families residing in a small township, so far as they can
be induced to attend it, even though only half their
members should be present at any one meeting. It
should limit speeches to ten minutes, excepting only
those addresses or essays which eminently qualified
persons are requested to specially prepare and read.
It should have a president, ready and able to repress
all ill-natured personalities, all irrelevant talk, and
especially all straying into the forbidden regions of
political or theological disputation. At each meeting,
the subject should be chosen for the next, and not
less than four members pledged to make some observations thereon, with liberty to read them if unused
to speaking in public. These having been heard,

the topic should be open to discussion by all present: the humblest and youngest being specially encouraged to state any facts within their knowledge
which they deem pertinent and cogent. Let every
person attending be thus incited to say something calculated to shed light on the subject, to say this in
the fewest words possible, and with the utmost care
not to annoy or offend others, and it is hardly possible that one evening per week devoted to these
meetings should not be spent with equal pleasure
and profit.

The chief end to be achieved through such meetings is a development of the faculty of observation
and the habit of reflection. Too many of us pass
through life essentially blind and deaf to the wonders
and glories manifest to clearer eyes all around us.
The magnificent phenomena of the Seasons, even
the awakening of Nature from death to life in
Spring-time, make little impression on their senses,
still less on their understandings. There are men
who have passed forty times through a forest, and
yet could not name, within half a dozen, the various
species of trees which compose it; and so with
everything else to which they are accustomed. They
need even more than knowledge an intellectual awakening; and this they could hardly fail to receive from
the discussions of an intelligent and earnest Farmers'
Club.

A genuine and lively interest in their vocation is
needed by many farmers, and by most farmers' sons.

Too many of these regard their homesteads as a prison, in which they must remain until some avenue of escape into the great world shall open before them. The farm to such is but the hollow log into which a bear crawls to wear out the rigors of Winter and await the advent of Spring. Too many of our boys fancy that they know too much for farmers, when in fact they know far too little. A good Farmers' Club, faithfully attended, would take this conceit out of them, imbuing them instead with a realizing sense of their ignorance and incompetency, and a hearty desire for practical wisdom.

A recording secretary, able to state in the fewest words each important suggestion or fact elicited in the course of an evening's discussion, would be hardly less valuable or less honored than a capable president. A single page would often suffice for all that deserves such record out of an evening's discussion ; and this, being transferred to a book and preserved, might be consulted with interest and profit throughout many succeeding years. No other duty should be required of the member who rendered this service, the correspondence of the Club being devolved upon another secretary. The habit of bringing grafts, or plants, or seeds, to Club meetings, for gratuitous distribution, has been found to increase the interest, and enlarge the attendance of those formerly indifferent. Almost every good farmer or gardener will sometimes have choice seeds or grafts to spare, which he does not care or cannot expect to sell, and these being distributed to the Club will not only increase its popularity, but give him a right to share when another's surplus is in like manner distributed. If one has choice fruits to give away, the Club will afford him an excellent opportunity; but I would rather not attract persons to its meetings by a prospect of having their appetites thus gratified at others' expense. A Flower-Show once in each year, and an Exhibition of Fruits and other choice products at an evening meeting in September or October, should suffice for festivals. Let each member consider himself pledged to˙bring to the Exhibition the best material result of his year's efforts, and the aggregate will be satisfactory and instructive.

The organization of a Farmers' Club is its chief difficulty. The larger number of those who ought to participate usually prefer to stand back, not committing themselves to the effort until after its success has been assured. To obviate this embarrassment, let a paper be circulated for signatures, pledging each signer to attend the introductory meeting and bring at least a part of his family. When forty have signed such a call, success will be well-nigh assured.

XLIV.

I HAVE already set forth my belief that Irrigation is everywhere practicable, is destined to be generally adopted, and to prove signally beneficent. I do not mean that every acre of the States this side of the Missouri will ever be thus supplied with water, but that *some* acres of every township, and of nearly every farm, should and will be. I propose herein to speak with direct reference to that large portion of our country which cannot be cultivated to any purpose without Irrigation. This region, which is practically rainless in Summer, may be roughly indicated as extending from the forks of the Platte westward, and as including all our present Territories, a portion of Western Texas, the entire State of Nevada, and at least nine-tenths of California. On this vast area, no rain of consequence falls between April and November, while its soil, parched by fervid, cloudless suns, and swept by intensely dry winds, is utterly divested of moisture to a depth of three or four feet; and I have seen the tree known as Buckeye growing in it, at least six inches in diameter, whereon every

(260)

leaf was withered and utterly dead before the end of August, though the tree still lived, and would renew its foliage next Spring.

Most of this broad area is usually spoken of as desert, because treeless, except on the slopes of its mountains, where certain evergreens would seem to dispense with moisture, and on the brink of infrequent and scanty streams, where the all but worthless Cotton-wood is often found growing luxuriantly. A very little low Gamma Grass on the Plains, some straggling Bunch-grass on the mountains, with an endless profusion of two poor shrubs, popularly known as Sage-brush and Grease-wood, compose the vegetation of nearly or quite a million square miles.

I will confine myself in this essay to the readiest means of irrigating the Plains, by which I mean the all but treeless plateau that stretches from the base of the Rocky Mountains, 300 to 400 miles eastward, sloping imperceptibly toward the Missouri, and drained by the affluents of the Platte, the Kansas, and the Arkansas rivers.

The North Platte has its sources in the western, as the South Platte has in the eastern, slopes of the Rocky Mountains. Each of them pursues a generally north-east course for some 300 miles, and then turns sharply to the eastward, uniting some 300 miles eastward of the mountains, where the Plains melt into the Prairies. Between these two rivers and the eastern base of the mountains lies an irregular delta or triangle, which seems susceptible of irrigation at a smaller cost than the residue. The location of Union Colony may be taken as a fair illustration of the process, and the facilities therefor afforded by nature.

Among the streams which, taking rise in the eastern gorges of the Rocky Mountains, run into the South Platte, the most considerable has somehow acquired the French name of Cache la Poudre. It heads in and about Long's Peak, and, after emerging from the mountains, runs some 20 to 25 miles nearly due east, with a descent in that distance of about 100 feet. Its waters are very low in Autumn and Winter, and highest in May, June and July, from the melting of snow and ice on the lofty mountains which feed it. Like all the streams of this region, it is broad and shallow, with its bed but three to four feet below the plains on either side.

Greeley, the nucleus of Union Colony, is located at the crossing of the Cache la Poudre by the Denver-Pacific Railroad, about midway of its course from the Kansas Pacific at Denver northward to the Union Pacific at Cheyenne. Here a village of some 400 to 500 houses has suddenly grown up during the past Summer.

The first irrigating canal of Union Colony leaves the Cache la Poudre six or eight miles above Greeley, on the south side, and is carried gradually further and further from the stream until it is fully a mile distant at the village, whence it is continued to the Platte. Branches or ditches lead thence northward, conveying rills through the streets of the village, the gardens or plats of its inhabitants, and the public square, or plaza, which is designed to be its chief ornament. Other branches lead to the farms and five-acre allotments whereby the village is surrounded; as still others will do in time to all the land between the canal and the river. In due time, another canal will be taken out from a point further up the stream, and will irrigate the lands of the colony lying south of the present canal, and which are meantime devoted to pasturage in common.

Taking the water out of the river is here a very simple matter. At the head of an island, a rude dam of brush and stones and earth is thrown across the bed of the stream, so as to raise the surface two or three feet when the water is lowest, and very much less when it is highest. Thus deflected, a portion of the water flows easily into the canal.

A very much larger and longer canal, leaving the Cache la Poudre close to the mountains, and gradually increasing its distance from that stream to four or five miles, is now in progress by sections, and is to be completed this Winter. Its length will be thirty miles, and it will irrigate, when the necessary sub-canals shall have been constructed, not less than 40,000 acres. But it may be ten years before all this work is completed or even required. The lands most easily watered from the main canal will be first brought into cultivation; the sub-canals will be dug as they shall be wanted.

At first, members of the Colony arriving at its location, hesitated to take farm allotments and build
upon them, from distrust of the capacities of the soil.
They saw nothing of value growing upon it; the little grass found upon it was short, thin, and brown. It
was not black, like the prairies and bottoms of Illinois
and Kansas, but of a light yellow snuff-color, and
deemed sterile by many. But a few took hold, and
planted and sowed resolutely; and, though it was too
late in the season for most grains, the results were
most satisfactory. Wheat sown in June produced 30
bushels to the acre; Oats did as well; while Potatoes, Beets, Turnips, Squashes, Cabbages, etc., yielded
bounteously; Tomatoes did likewise, but the plants
were obtained from Denver. Little was done with
Indian Corn, but that little turned out well, though I
judge that the Summer nights are too cold here to
justify sanguine expectations of a Corn-crop—the
altitude being 5,000 feet above the sea, with snow-
covered mountains always visible in the west. For
other Grains, and for all Vegetables and Grasses, I
believe there is no better soil in the world.

To many, the cost of Irrigation would seem so
much added to the expense of cultivating without
irrigation; but this is a mistake. Here is land entirely free from stump, or stick, or stone, which may
easily and surely be plowed or seeded in March or
April, and which will produce great crops of nearly
every grain, grass or vegetable, with a very moderate
outlay of labor to subdue and till it. The farmer

need not lose three days per annum by rains in the growing season, and need not fear storm or shower when he seeks to harvest his grass or grain. Nothing like ague or any malarious disease exhausts his vitality or paralyzes his strength. I saw men breaking up for the first time tracts which had received no water, using but a single span of horses as team; whereas, breaking up in the Prairie States involves a much larger outlay of power. The advantage of early sowing is very great; that of a long planting season hardly less so. I believe a farmer in this colony may keep his plow running through October, November, and a good portion of December; start it again by the 1st of March, and commence seeding with Wheat, Oats, and Barley, and keep seeding, including planting and gardening, until the first of June, which is soon enough to plant potatoes for Winter use. Thenceforth, he may keep the weeds out of his Corn, Roots, and Vegetables, for six weeks or two months; and, as every day is a bright working-day, he can get on much faster than he could if liable to frequent interruptions by rains. I estimate the cost of bringing water to each farm at $5 per acre, and that of leading it about in sub-ditches, so that it shall be available and applicable on every acre of that farm, at somewhat less; but let us suppose that the first cost of having water everywhere and always at command is $10 per acre, and that it will cost thereafter $1 per acre to apply it, I maintain that it is richly worth having, and that nearly every farm

product can be grown cheaper by its help than on
lands where irrigation is presumed unnecessary.
There are not many acres laid down to grass in New-
England, whether for hay or pasture, that would not
have justified an outlay of $10 per acre to secure
their thorough irrigation simply for this year alone.

XLV.

SEWAGE.

THE great empires of antiquity were doomed to
certain decay and dissolution by a radical vice inherent in their political and social constitution.   Power
rapidly built up a great capital, whereto population
was attracted from every quarter; and that capital
became a focus of luxury and consumption.   Grain,
Meat, and Vegetables—the fat of the land and the
spoils of the sea—were constantly absorbed by it in
enormous quantities ; while nothing, or, at best, very
little, was returned therefrom to the continually ex-
hansted and impoverished soil.   Thus, a few ages, or
at most a few centuries, sufficed to divest a vast surrounding district, first, of its fertility, ultimately of
its capacity for production.   And so Nineveh, Thebes,
Babylon, successively ceased to be capitals, and became ruins amid deserts.   Rome impoverished Italy

south of the Apennines; then Sicily; and, at last, Egypt: her sceptre finally departing, because her millions could no longer be fed without dispersion.

That some means must be devised whereby to return to the soil those elements which the removal of crop after crop inevitably exhausts, is a truth which has but recently begun to be clearly understood. Unluckily, the difficulty of such restoration is seriously augmented by the fact that cities, and all considerable aggregations of human beings, tend strongly in our day to locations by the sea-side, in valleys, and by the margins of rivers. Anciently, cities and villages were often built on hill-tops, or at considerable elevations, because foes could be excluded or repelled from such locations more surely, and with smaller force, than elsewhere. From such elevations, it need not have been difficult to diffuse, by means of water, all that could be gladly spared which would aid to fertilize the adjacent farms and gardens. A kindred distribution of the exuviæ of our modern cities is a far more difficult and costly undertaking, and involves bold and skillful engineering.

Yet the problem, though difficult, must be solved, or our great cities will be destroyed by their own physical impurities. The growth and expansion of cities, throughout the present century, have been wholly beyond precedent; and thus the difficulty of making a satisfactory disposition of their offal has been fearfully augmented. The sewerage of our streets and houses modifies the problem, but does not solve it. Desolating epidemics, like the Plague, Yellow Fever, and the Cholera, will often visit our great cities, and decimate their people, unless means can be found to cleanse them wholly and incessantly of whatever tends to pollute and render noisome their atmosphere.

SEWAGE is the term used in England to designate water which, having been slightly impregnated with the feculence and ordure of a city or village, is diffused over a farm or farms adjacent, in order to impart at once fertility and moisture to its soil. To secure an equable and thorough dissemination of Sewage, it is essential that the land to which it is applied, if not originally level or nearly so, shall be brought into such condition that the impregnated water may be applied to its entire surface, and shall thence settle into, moisten, and fertilize, each cubic inch of the soil. This involves a very considerable initial outlay; but the luxuriance of the crops unfailingly produced, under the influence of this vivifying irrigation, abundantly justifies and rewards that outlay.

As yet, the application of Sewage is in its infancy; since the perfect and total conversion of all that a great city excretes into the most available food for plants, requires not only immense mains and reservoirs, with a costly network of distributing dykes or ditches, but novel appliances in engineering, and a large investment of time as well as money. Years must yet elapse before all the excretions of a great

city like London or New-York can thus be trans-
muted into the means of fertilizing whole counties
in their vicinity. But the work is already well begun, and another generation will see it all but completed. Meantime, many smaller cities, more eligibly
located for the purpose, are already enriching by
their Sewage the rural districts adjacent, which they
had previously tended strongly to impoverish. Edin-
burgh, the capital of Scotland, is among them. The
little village of Romford, England, is one of those
which have recently been made to contribute by
Sewage to this beneficent end; and a visit of inspection paid to it, on the 15th of October last, by the
London Board of Works, elicited accounts of the
process and its results, in the London journals, which
afforded hints for and incitement to similar undertakings in this and other countries—undertakings
which may be postponed, but the only question is
one of time. *The Daily News* of Oct. 17th, says:

"Breton's Farm consists of 121 acres of light and
poor gravelly soil; and it now receives the whole
available sewage of the town of Romford—that is, of
about 7,000 persons. This is conveyed to the land
by an iron pipe of 18 inches in diameter, which is
laid under ground, and discharges its contents into
an open tank. From this tank, the sewage is pumped to a height of 20 feet, and is then distributed over
the land by iron or concrete troughs, or 'carriers,'
fitted with sluices and taps, so that the amount of
sewage applied to any given portion of the field can

be regulated with the greatest facility and nicety. To insure the regular and even flow of the sewage when discharged from the carriers, it was necessary to lay out the land with mathematical accuracy; and it has been leveled and formed by the theodolite into rectilinear beds of uniform width of thirty feet, slightly inclining from the centres, along which the sewage is applied. The carriers or open troughs, by which the sewage is conveyed, run along the top of each series of these beds or strikes; and at the bottom there is in every case a good road, by means of which free access is provided for a horse and cart, or for the steam plow—the use of which is in contemplation—to every bed and crop. These arrangements—the carrying out of which involved the removal of six hundred trees and a great length of heavy fences, the filling up of a number of ditches and no less than nine ponds, as well as the complete under-draining of the whole farm—were mainly effected last year; but it was not until the middle of April, 1870, that Mr. Hope received any of this sewage from the town of Romford, and not until the following month that he obtained both the day and night supply. Satisfactory, therefore, as have been the results of the present season's operations, they have been obtained under disadvantageous circumstances, and cannot be regarded as affording complete evidence of the benefits which may be derived from the application of sewage to even a poor and thin soil, which had already ruined more than one of those who had attempted to cultivate it. To mention only one drawback which arose from the lateness of the period at which the sewage was first received, Mr. Hope had not the advantage of being able to apply it to his seed-beds: and thus many, if not all his plants were not ready for setting out so early as they would be in a future year, and some of the crops have suffered in consequence—that is to say, have suffered in a comparative sense. Speaking positively, they have in all instances been much larger, not only than any that could have been grown upon the same land without the use of sewage, but than any which have been raised from much superior land in the immediate neighborhood. The crops which have been or are being raised on different parts of the farm, are of diverse character; but, with all, the method of cultivation adopted has been attended with almost equal success. Italian rye-grass, beans, peas, mangolds, carrots, broccoli, cabbages, savoys, beet-root, Batavia yams, Jersey cabbages, and Indian corn, have all grown with wonderful rapidity and yielded abundant harvests under the stimulating and nourishing influence of the Romford sewage. The visitors of Saturday last, as they tramped. over the farm under the guidance of its energetic proprietor, had an opportunity of witnessing the abundance and excellence of many of these crops. Even where the mangolds, from being planted late, had not attained any extraordinary size, it was noticeable that the plants were especially vigorous, and that there was not a vacant space in any of the rows. All the plants which had been placed in the ground had thriven, and would give a good return. Where this crop had been specially treated with a view to forthcoming shows, the roots had attained an enormous size, and, like some of the cabbages, had assumed almost gigantic proportions. The carrots were very fine and well-grown, and the heads of the Walcheren broccoli were as white, and firm, and crispy, as the finest cauliflowers; while the savoys, of unusual size and weight, were as round and hard as cannon balls; and some of the drumhead cabbages, although equally distinguished for closeness and firmness, were large enough in the heart to hold a good-sized child, and might, as was suggested upon the ground, very well be introduced into some pantomimic scene representing the kingdom of Brobdignag. The Indian corn had reached the respectable height of some eight feet, and, with few exceptions, each stalk carried a good-sized and well-filled cob or ear. These, unless we should have another spell of exceptionally hot weather, will not ripen; but in their green state they are readily eaten by horses and cattle, and prove excellent fodder.

In the course of their peregrinations, Mr. Hope's guests of course paid a visit to the tank in which the sewage is received before it is pumped on the land. We need hardly say that the appearance of this miniature lake of nastiness was anything but agreeable; but its odor was by no means overpowering, nor, indeed, very offensive. The rill of bright, clear water which flowed in at one corner, and some of which was handed about in tumblers, looking as pure as the limpid stream which flows from the most effective filters that are to be seen in the windows of London dealers, had only a short time before flowed out of this hideous reservoir in a very different state. We had met it in the "carriers" flowing along in a dark, inky stream, not smelling much, but covered with an ugly gray froth which reminded one of some of the most disagreeable details in the manufacture of sugar and rum, or suggested the idea that it had been used for a very foul wash indeed. With these reminiscences fresh in one's memory, it required some courage to comply with the pressing invitations to taste this 'effluent water.' There were, however, many of the party who braved the attempt; and, by all who tasted it, the water was pronounced to be destitute of any except a slightly mineral flavor. In dry weather, this effluent water, which has passed through the land and been collected by the drains, after mixing with the sewage, is again pumped over the fields; in wet weather, it can be turned into the brook which is dignified with the name of the river Rom. * * * We have omitted to mention that the rent paid by Mr. Hope is £3 per acre, and the cost of the sewage (at 2s. per head) £6 more."

—I think few thoughtful readers will doubt that here is the germ of a great movement in advance for the Agriculture of all old and densely peopled

12*

communities, and that our youngest cities and manufacturing villages may wisely consider it deeply,
with a view to its ultimate if not early imitation.
That we are not prepared to incur the inevitable expense of a thorough system of sewerage with reference
to the application to the soil of all the fertilizing
elements that a city would gladly spare, by no means
proves that we should not consider and plan with a
view to the ultimate creation and utilization of
Sewage.

XLVI.

MORE OF IRRIGATION.

I have thus far considered Irrigation with special
reference to those limited, yet very considerable districts, which are traversed or bordered by living
streams, and, having a level or slightly rolling surface, present obvious facilities for and incitements
to the operation. Such are the valleys of the Platte,
and of nearly or quite all its affluents after they leave
the Rocky Mountains; such is the valley of the upper
Arkansas; such the valleys of the Smoky Hill and the
Republican, so far down as Irrigation may· be considered necessary. Irrigation on all these seems to me
inevitable, and certain to be speedily, though capri-
ciously, effected.

I believe a dam across either fork of the Platte, at

any favorable point above their junction, raising the surface of the stream six feet, at a cost not exceeding $10,000, would suffice to irrigate completely not less than fifty square miles of the valley below it, while serving at the same time to furnish power for mills and factories to a very considerable extent; for the need of Irrigation is not incessant, but generally confined to two or three months per annum, and all of the volume of the stream not needed for Irrigation could be utilized as power. Thus the valleys of the few constant water-courses of the Plains may come at an early day to employ and subsist a dense and energetic population, engaged in the successful prosecution alike of agriculture and manufactures, while belts, groves, and forests, of choice, luxuriant timber, will diversify·and embellish regions now bare of trees, and but thinly covered with dead herbage from June until the following April.

But, when we rise above the bluffs, and look off across the blank, bleak areas where no living water exists, the problem becomes more difficult, and its solution will doubtless be much longer postponed. To a stranger, these bleak uplands seem sterile; and, though such is not generally the fact, the presumption will repel experiments which involve a large initial outlay. The railroad companies, which now own large tracts of these lands, will be obliged either to demonstrate their value, or to incite individuals and colonists to do it by liberal concessions. As the case stands to-day, most of these lands, which would have been dear at five cents per acre before the roads were built, could not be sold at any price to actual settlers; even with the railroad in plain sight, because of the dearth of fuel and timber, and because also the means of rendering them fruitful and their cultivation profitable are out of reach of the ordinary pioneer. Hence, so long as the valleys of the living streams proffer such obvious invitations to settlement and tillage, by the aid of Irrigation, I judge that the higher and dryer plains will mainly be left to the half-savage herdsmen who rear cattle and sheep without feeding and sheltering them, by giving them the range of a quarter-section to each bullock, and submitting to the loss of a hundred head or so after each great and cold snow-storm, as an unavoidable dispensation of Providence.

But in process of time even the wild herdsmen will be softened into or replaced by regular farmers, plowing and seeding for vegetables and small grains, sheltering their habitations with trees, and sending their children to school. This change involves Irrigation; and the following are among the ways in which it will be effected:

The Plains are nowhere absolutely flat (as I presume the "desert" of Sahara is not), but diversified by slopes, and swells, and gentle ridges or divides, affording abundant facilities for the distribution of water. A well, sunk on the crest of one of these divides, will be filled with living water at a depth ranging from 50 to 100 feet. A windmill of modest

dimensions placed over this well will be rarely stopped for want of impelling power: Wind being,
next to space, the thing most abundant on the
Plains. A reservoir or pond covering three or four
acres may be made adjacent to the well at a small
cost of labor, by excavating slightly and using the
earth to form an embankment on the lower side.
The windmill, left alone, will fill the reservoir during
the windy Winter and Spring months with water
soon warmed in the sun, and ready to be drawn off
as wanted throughout the thirsty season of vegetable
growth and maturity. Carefully saved, the product
of one well will serve to moisten and vivify a good
many acres of grass or tillage.

Such is the retail plan applicable to the wants of
solitary farmers; but I hope to see it supplemented
and invigorated by the extensive introduction of
Artesian wells, whereof two, by way of experiment,
are now in progress at Denver and Kit Carson respectively.

I need not here describe the Artesian well, farther
than to say that it is made by boring to a depth
ranging from 700 to more than a 1000 feet, tubing regularly from the top downward until a stream is
reached which will rise to and above the surface,
flowing over the top of the tube in a stream often as
large as an average stove-pipe. Such a well, after
supplying a settlement or modest village with water,
may be made to fill a reservoir that will sufficiently
irrigate a thousand cultivated acres. Its water will

usually be warmer than though obtained from near the surface, and hence better adapted to Irrigation.

Of course, the Artesian well is costly, and will not soon be constructed for uses purely agricultural; but the railroads traversing the Plains and the Great Basin will sometimes be compelled to resort to one without having use for a twentieth part of the water they thus entice from the bowels of the earth; and that which they cannot use they will be glad to sell for a moderate price, thus creating oases of verdure and bounteous production. The palpable interest of railroads in dotting their long lines of desolation with such cheering contrasts of field and meadow and waving trees, render nowise doubtful their hearty coöperation with any enterprising pioneer who shall bring the requisite capital, energy, knowledge, and faith, to the prosecution of the work.

These are but hasty suggestions of methods which will doubtless be multiplied, varied, and improved upon, in the light of future experience and study. And when the very best and most effective methods of subduing the Plains to the uses of civilized man shall have been discovered and adopted, there will still remain vast areas as free commons for the herdsmen and sporting-grounds for the hunter of the Elk and the Antelope, after the Buffalo shall have utterly disappeared.

I do not doubt the assertion of the plainsmen that rain increases as settlements are multiplied. Crossing the Plains in 1859, I noted indications that timber had formerly abounded where none now grows; and I presume that, as young trees are multiplied in the wake of civilization, finally thickening into clumps of timber and beginning a forest, more rain will fall, and the extension of woodlands become comparatively easy. But, relatively to the country eastward of the Missouri, the Plains will always be arid and thirsty, with a pure, bracing atmosphere that will form a chief attraction to thousands suffering from or threatened with pulmonary afflictions. A million of square miles, whereon is found no single swamp or bog, and not one lake that withstands the drouth of Summer, can never have a moist climate, and never fail to realize the need of Irrigation.

The Plains will in time give lessons, which even the well-watered and verdurous East may read with profit. Such level and thirsty clays as largely border Lake Champlain, for example, traversed by streams from mountain ranges on either hand, will not always be owned and cultivated by men insensible to the profit of Irrigation. Nor will such rich valleys as those of the Connecticut, the Kennebec, the Susquehanna, be left to suffer year after year from drouth, while the water which should refresh them runs idly and uselessly by. Agriculture repels innovation, and loves the beaten track; but such lessons as New-England has received in the great drouth of 1870 will not always be given and endured in vain.

XLVII.

THE more I consider the present state of our Agri culture, the more emphatic is my discontent with the farmer's present sources and command of power. The subjugation and tillage of a farm, liko the running of a factory or furnace, involves a continual use of Power; but the manufacturer obtains his from sources which supply it cheaply and in great abundance, while the farmer has been content with an inferior article, in limited supply, at a far heavier cost. Yet the stream which turns the factory's wheels and sets all its machinery in motion traverses or skirts many farms as well, and, if properly harnessed, is just as ready to speed the plow as to impel the shuttles of a woolen-mill, or revolve the cylinders of a calico-printery. Nature is impartially kind to all her children; but some of them know how to profit by her good-will far more than others. No doubt, we all have much yet to learn, and our grandchildren will marvel at the proofs of stupidity evinced in our highest achievements; but I am not mistaken in asserting that, as yet, the farmers' control of Nature's free gifts of power is very far inferior to that of nearly every other class of producers.

I have been having much plowing done this Fall—in my orchards, for what I presume to be the good of the trees; on my drained swamp, because it is not yet fully subdued and sweetened, and I judge that the Winter's freezing and thawing will aid to bring it into condition. And then my swamp lies so low and absolutely flat that the thaws and rains of Spring render plowing it in season for Oats, or any other crop that requires early seeding, a matter of doubt and difficulty. All the land I now cultivate, or seek to cultivate, has already been well plowed more than once; no stump or stone impedes progress in the tracts I have plowed this Fall; yet a good plow, drawn by two strong yoke of oxen, rarely breaks up half an acre per day; and I estimate two acres per week about what has been averaged, at a cost of $18 for the plowman and driver; offsetting the oxen's labor against the work done by the men at the barn and elsewhere apart from plowing. In other words: I am confident that my plowing has cost me, from first to last, at least, $10 per acre, and would have cost still more if it had been done as thoroughly as it ought. I am quite aware that this is high—that sandy soils and dry loams are plowed much cheaper; and that farmers who plow well (with whom I do not rank those who scratch the earth to a depth of four or five inches) do it at a much lower rate. Still, I estimate the average cost in this country of plowing land twelve inches deep at $5 per acre ; and I am confident that it does not cost one cent less.

Nor is cost the only discouragement. There is not half so much nor so thorough plowing among us, especially in the Fall, as there should be. The soil is, for a good part of the time, too dry or too wet; the weather is inclement, or the ground is frozen: so the plow must stand still. At length, the signs are auspicious; the ground is in just the right condition; and we would gladly plow ten, twenty, fifty acres during the brief period wherein it remains so; but this is impossible. Others want to improve the opportunity as well as we ; extra teams are rarely to be had at any price; and our own slow-moving oxen refuse to be hurried. Standing half a mile off, you *can* see them move, if your eye-sight is keen, and you have some stationary object interposed whereby to take an observation; but it is as much as ever. If your soil is such that you can use horses, you get on, of course, much faster; but all that you gain in breadth you are apt to lose in depth. There may be spans that will take the plow right along though you sink it to the beam; but they are sure to be slow travelers. I never knew a span that would plow an acre per day as I think it should be plowed ; though, if your only object be to get over as much ground as possible, you may afflict and titillate two acres, or as much more as you please.

Now, I have before me a letter to *The Times* (London) by Mr. William Smith, of Woolston, Bucks, who states that he has just harvested his fifteenth annual crop cultivated by steam-power, and has prepared his land for the sixteenth ; and he gives details, showing that he breaks up and ridges heavy clay soils at the rate of six acres per day, and plows lands already in tillage at the rate of fully nine acres per day. He gives the total cost, (including wear and tear,) of breaking up a foot deep and ridging 65 acres in September and October in this year, 1870, at £20 6s. 6d. or about $100 in gold: call it $112 in our greenbacks, and still it falls consideraby below $2 (greenbacks) per acre. Say that labor and fuel are twice as dear in this country as in England, and this would make the cost of thoroughly pulverizing by steam-power a heavy clay soil to a depth of twelve inches less than $4 per acre here. I do not believe this could be done by animal power at $10 per acre, not considering the difficulty of getting it thoroughly done at all. Mr. Smith pertinently says: " Horse-power could not give at any cost such valuable work as this steam-power ridging and subsoiling is." He tills 166 acres in all, making the cost of steam-plowing his stubble-land 4s. 8½d. per acre (say $1 30 greenback). And he gives this interesting item :

" No. 5, light land, 12 acres, was ridge-plowed and subsoiled last year for beans : that operation left the land, after the bean-crop came off, in so nice a state, that cultivating once over with horses, at a cost of 2s. per acre, was all that was needed this Autumn for wheat next year. The wheat was drilled four days back."

—Now I am not commending Steam as the best source of power in aid of Agriculture. I hope we shall be able to do better ere long. I recognize the enormous waste involved in the movement of an engine, boiler, etc., weighing several tons, back and forth across our fields, and apprehend that it must be difficult to avoid a compression of the soil therefrom. A stationary engine and boiler at either end of the field, hauling a gang of plows this way and that by means of ropes and pulleys, must involve a very heavy outlay for machinery and a considerable cost in its removal from farm to farm, or even from field to field. Either of these may be the best device yet perfected; but we are bound to do better in time.

Precisely how and when the winds which sweep over our fields shall be employed to pulverize and till the soil, are among the many things I do not know; but, that the end will yet be achieved, I undoubtingly trust. I know somewhat—not much—of what has been done and is doing, both in Europe and America, to extend and diversify the utilization of wind as a source of power, and to compress and retain it so that the gale which sweeps over a farm to-night may afford a reserve or fund of power for its cultivation on the morrow or thereafter. I know a little of what has been devised and done toward converting and transmitting, through the medium of compressed air, the power generated by a waterfall—say Niagara or Minnehaha—so that it may be expended and utilized at a distance of miles from its source, impelling machinery of all kinds at half the cost of steam. I know vaguely of what is being done with Electricity, with an eye to its employment in the production of power, by means of enginery not a tenth so weighty and cumbrous as that required for the generation and utilization of Steam, and by means of a consumption (that is, transformation) of materials not a hundredth part so bulky and heavy as the water and steam which fill the boilers of our factories and locomotives. I am no mechanician, and will not even guess from what source, through what agencies, the new power will be vouchsafed us which is in time to pulverize our fields to any required depth with a rapidity, perfection, and economy, not now anticipated by the great body of our farmers. But my faith in its achievement is undoubting; and, though I may not live to see it, I predict that there are readers of this essay who will find the forces abundantly generated all around us by the spontaneous movement of Wind, Water, and Electricity—one or more, and probably by all of

ized

the farmer's labor, while quadrupling its efficiency in producing all by which our Earth ministers to the sustenance and comfort of man.

XLVIII.

COMPLAINT is widely made of a decrease in the relative population of our rural districts; and not without reason, or, at least, plausibility. I presume the Census of 1870 will return no more farmers in the State of New York, and probably some fewer in New England, than were shown by the Census of 1860. The very considerable augmentation of the number of their people will be found living wholly in the cities and incorporated villages. I doubt whether there are more farmers in the State of New York to-day than there were .in 1840, though the total population has meantime doubled. Many farms have been transformed into country-seats for city bankers, merchants, and lawyers; others have been consolidated, so that what were formerly two or three, now constitute but one; and, though every body says, "Our farms are too large for our capital," "We run over too much land," etc., etc., yet, I can hear of few farms that have been, or are expected to be, divided, except into village or city lots; while the prevalent tendency is still the other way. An ineffi-

(286)

cient farmer dies heavily in debt, or is sold out by the sheriff: his farm is rarely divided between two purchasers, while it is quite often absorbed into the estate of some thrifty neighbor; and thus small farmers are selling out and moving westward much oftener than large ones. Such are the obvious facts: now for some of the reasons ·

I. Our State, like New England, was originally all but covered by a heavy growth of forest. The removal of this timber involved very much hard work, most of which has been done in this century, and much of it by the present generation. When I first traversed Chautauqua County, forty-three years ago, from two-thirds to three-fourths of her acres must have been still covered with the primeval forest—a tall, heavy growth of Beech, Maple, Hemlock, White Pine, etc., which yielded very slowly to the efforts of the average chopper. Many a pioneer gave half his working hours for twenty years to the clearing off of Timber, Fencing, cutting out roads, etc., and had not sixty acres in arable condition at the last. Outside of the villages, the population of that county was probably as great in 1830 as it is to-day, though the annual production of her tillage was not half what it now is. Her farms are now made; her remaining wood-lands are worth about as much per acre as her tillage; there is now comparatively little timber-cutting, or land-clearing; and two-thirds of the pioneers, or their sons who inherited their farms, have sold out, or *been* sold out, and pushed further westward.

Meantime, Grazing and Dairying have extensively supplanted Grain-growing; and farmers who found more work than they could do on 60 or 80 acres, now manage 160 to 320 acres with ease. I do not say that they ought not to farm better; I only state the facts that they thrive by this dairy-farming, and are not exhausting their lands. And what is true of Chautauqua is measurably true of half the rural Counties in our State.

II. Formerly, Wood was the only fuel known to our farmers, while immense quantities of it were burned in our cities, at the salt-works, etc. At present, wood is scarcely used for fuel, except as kindling, in any of our cities, villages, or manufactories, while the consumption of Coal by our farmers is already very large, and rapidly extending. All this reduces the demand for labor on our farms and in our forests, while increasing the corresponding demand in the Coal Mines, and on the railroads. Luzerne County, Pennsylvania, has doubled her population within the last twelve or fourteen years; and this at the expense of our rural districts.

III. Our agricultural implements and machinery grow annually more effective, and at the same time more costly. The outfit of a good farm costs five-fold what it did forty years ago. The farmer makes and secures his Hay far more rapidly and effectively than his father did, but pays far more for Reapers, Mowers, Rakers, etc.; in other words, he makes Winter work abridge that of Summer—makes a huudred days' work in some village or city save thrice as many days' work on his farm. This enhances his profits, but swells our urban, while it diminishes our rural population.

IV. Much has been said of the degeneracy and increasing sterility of the New England Puritan stock. All this is shallow and absurd. There never before were so many people who proudly traced their origin to a New England ancestry as now. What is true in the premises is this: The New England stock is becoming very widely diffused, and is giving place, to a considerable extent, to other elements in its original home. Forty years ago, at least seven-eighths of the inhabitants of Boston were of New England birth and lineage; now, hardly half are so. The descendants of the Pilgrims are scattered all over our wide country; while hundreds of thousands have flowed in from Ireland, from Germany, from Canada, to fill the places thus relinquished; and, since most of the immigrants, whether into or out of New England, seek their future homes in the spring-time of life, their children are mainly born to them after rather than before their migration. The Yankees have no fewer children than formerly; but they are now born in Minnesota, in Illinois, in Kansas; while those born in New England are, for identical reasons, in large proportion of Irish or of Canadian parentage. There are New England townships, whereof most of the heads of families are long past the prime of life; their children having left them

13

for more attractive localities, and the work on their farms being now done mainly by foreign-born employés. As a general rule, the boys first wandered off, leaving the girls only the alternative of following, or dying in maidenhood. Marked diversities of race, of creed, and of education, have thus far prevented any considerable intermingling of the Yankee with the foreign element by marriage. And what is true of New England is measurably true of our own State.

I have not intended by these observations to combat the assumption that our people too generally prefer other employments to farming. The obstacles to effective modern Agriculture—that is, to agriculture prosecuted by the help of efficient machinery— presented by that incessant alternation of rock and bog, which characterizes New England and some parts of New York, I have already noted; and they interpose a serious, discouraging impediment to agricultural progress. A farm intersected by two or three swamps and brooks, separated by steep, rocky ridges, and dotted over with pebbly knolls, sometimes giving place to a strip of sterile sand, is far more repulsive to the capable, intelligent farmer of to-day than it was to his grandfather. So far as my observation extends, there are more New England farms on which you cannot, than on which you can, find ten acres in one unbroken area suitable for planting to Corn, or sowing to Winter Grain. Hence, Agriculture in the East will always seem petty and irregular when brought into contrast with the prairie cultivation of the West. Grain can never be grown here so cheaply nor so abundantly as there; while the tendency of our pastures to cover themselves over with moss and worthless shrubs, unless frequently broken up and rëseeded, makes even dairying more difficult and costly in New England and along its western border than in almost any other part of our country.

Yet, these discouragements are balanced by compensations. Timber springs luxuriantly and grows rapidly throughout this region; while our harsh, capricious climate gives to our Hickory, White Oak, White Ash, and other varieties, qualities unknown to such grown elsewhere, while prized everywhere. Apples, and most fruits of the Temperate Zone, do well with us; while our cities and manufacturing villages proffer most capacious markets. Potatoes and other edible roots produce liberally, and generally command good prices. Hay sells for $12 to $30 per ton, is easily grown, and is in eager and increasing demand. We ought to produce twice our present crop from the same area, and have need of every pound of it; for neither our cattle nor our sheep are nearly so numerous nor so well fed as they should be. In short, there is money to be made, by those who have means and know how, by buying New England farms, tilling them better, and growing much larger crops than their present occupants have done. There are many who can do better in the West; but the

right men can still make money by farming this side
of the Susquehanna and the Genesee; and I would
gladly incite some thousands more of them to try.

XLIX.

LARGE AND SMALL FARMS.

THERE is fascination for most minds in naked magnitude. The young colonel, who can hardly handle
a brigade effectively in battle, would like of all things
to command a great army; and the tiller of fifty
rugged acres has his ravishing dreams of the delights
inherent in a great Western farm, with its square
miles of corn-fields, and its thousands of cattle. Each
of them is partly right and partly wrong.

There are generals capable of commanding 100,000
men. Napoleon says there were two such in his day
—himself and another: and these generally find the
work they are fit for, without special effort or aspiration. So there are men, each of whom can really
farm a township, not merely let a herd of cattle
roam over it unfed and unsheltered, living and dying
as may chance: the owners expecting to grow rich
by their natural increase. This *ranching* is not
properly farming at all, but a very different and far
ruder art. I judge that the farmers who can really

till—or even graze—several thousand acres of land, so as to realize a fair interest on its value, are even scarcer than the farms so capacious.

But there is such a thing as farming on a large scale; and it is a good business for those who understand it, and have all the means it requires. The farmer who annually grows a thousand acres of good Grain, and takes reasonable care of a thousand head of Cattle, is to be held in all honor. He will usually grow both his Grain and his Beef cheaper than a small farmer could do it, and will generally find a good balance on the right side when he makes up and squares his accounts of a year's operations. I could recommend no man to run into debt for a great farm, expecting that farm to work him out of it; but he who inherited or has acquired a large farm, well stocked, and knows how to make it pay, may well cling to it, and count himself fortunate in its possession. But the great farmer is already regarded with sufficient envy. Most boys would gladly be such as he is; the difficulty in the case is that they lack the energy, persistency, resolution, and self-denial, requisite for its achievement.

We will leave large farms and farming to recommend themselves, while we consider more directly the opportunities and reasonable expectations of the small farmer.

The impression widely current that money cannot be made on a small farm—that, in farming, the great fish eat up the little ones—is deduced from very im

perfect data. I have admitted that Grain and Beef
can usually be produced at less cost on great than on
small farms, though the rule is not without exceptions. I only insist that there are room and hope
for the small farmer also, and that large farming can
never absorb nor enable us to dispense with small
farms.

I. And first with regard to Fruit. Some Tree
Fruits, as well as Grapes, are grown on a large scale
in California—it is said, with profit. But nearly all
our Pears, Apples, Cherries, Plums, etc., are grown
by small farmers or gardeners, and are not likely to
be grown otherwise. All of them need at particular
seasons a personal attention and a vigilance which
can seldom or never be accorded by the owners or
renters of large farms. Should small farms be generally absorbed into larger, our Fruit-culture would
thenceforth steadily decline.

II. The same is even more true of the production
of Eggs and the rearing of Fowls. I have had knowledge of several attempts at producing Eggs and
Fowls on a large scale in this country, but I have no
trustworthy account of a single decided success in
such an enterprise. On the contrary, many attempts
to multiply Fowls by thousands have broken down,
just when their success seemed secure. Some con-
tagious disease, some unforeseen disaster, blasted the
sanguine expectations of the experimenter, and trans-
muted his gold into dross.

Yet, I judge that there is no industry more capable

of indefinite extension, with fair returns, than Fowl-breeding on a moderate scale. Eggs and Chickens are in universal demand. They are luxuries appreciated alike by rich and poor; and they might be doubled in quantity without materially depressing the market. Our thronged and fashionable watering-places are never adequately supplied with them; our cities habitually take all they can get and look around for more. I believe that twice the largest number of Chickens ever yet produced in one year might be reared in 1871, with profit to the breeders. Even if others should fail, the home market found in each family would prove signally elastic.

This industry should especially commend itself to poor widows, struggling to retain and rear their children in frugal independence. A widow who, in the neighborhood of a city or of a manufacturing village, can rent a cottage with half an acre of south-ward-sloping, sunny land, which she may fence so tightly as to confine her Hens therein, whenever their roaming abroad would injure or annoy her neighbors, and who can incur the expense of constructing there-on a warm, commodious Hen-house, may almost certainly make the production of Eggs and Fowls a source of continuous profit. If she can obtain cheaply the refuse of a slaughter-house for feed, giving with it meal or grain in moderate quantities, and according that constant, personal, intelligent supervision, without which Fowl-breeding rarely prospers, she may reasonably expect it to pay, while affording her

an occupation not subject to the caprices of an em-
ployer, and not requiring her to spend her days away
from home.

III. Though the ordinary Market Vegetables may
be grown on large farms, the fact that they seldom
are is significant.   Cabbages, Peas, Poled Beans,
Tomatoes, and even Potatoes, are mainly grown on
small farms, as they always have been.   There are
sections wherein no cash market for Vegetables exists or can be relied on ; and here they will continue
to be grown to the extent only of the growers' respective needs ; but wherever the prevalence of manufactures or the neighborhood of a great city gives
reasonable assurance of a market, they are grown at
a profit per acre which is rarely realized from a
Grain-crop.   No less than $100 per acre is often, if
not generally, achieved by the growers of Cabbage
around this city ; and this not from rich, deep
garden-mold, but from fair farming land, underdrained, subsoiled, and liberally manured.

The careless, slipshod farmer may do better—that
is, he will not fail so signally—in Grain cultivation ;
but there are few more decided or brilliant successes
than have been achieved within the last few years
within sight of this City, and wholly in the tillage of
small farms.

I trust I have here said enough to show that there
is a legitimate and promising field for agricultural
enterprise and effort, other than that which contem-

plates the acquisition and rule of a township, and that, while farming on a large area is to many attractive and inspiring, there are scope and incitement also for tillage on a humbler scale—for tillage that permits no weed to ripen seed, and no nest of caterpillars to flourish a month undisturbed—for tillage that achieves large crops and profits from small areas, and rejoices in that neatness and perfection of culture attainable only in the management of small farms.

L.

EXCHANGE AND DISTRIBUTION.

THE machinery whereby the farmer of our day converts into cash or other values that portion of his products which is not consumed in his house or on his farm, seems to me lamentably imperfect. Let me illustrate my meaning:

After three all but fruitless years, we have this year a bountiful Apple-crop, in this State and (I believe) throughout the North. Our old orchards being still, for the most part, preserved and in bearing condition, while a good many young ones, planted ten to twenty years ago, begin to fruit considerably, we had, throughout the three Fall months, a superabundance of this homely, wholesome, palatable fruit. It should have been cheap for the great body of our mechanics and laborers to provide their families with all the ripe, good Apples that they could consume without injuring themselves by gluttony. Good Apples should have been constantly displayed on every workingman's table, to be eaten raw as a dessert, or baked and eaten with bread and milk for breakfast or supper. Each provident housewife should now have her tub of apple-sauce, her barrel of dried apples, or both, for Winter use; while a dozen bushels of good keepers should be stored in every cellar, to be drawn upon from day to day during the next four or five months. In short, Apples should have been and be, from last August to next May, as common as bread and potatoes, and should have been and be as freely eaten in every household and by every fireside.

How nearly have we realized this?

I will not guess how many millions of bushels have rotted under the trees that bore them, been eaten by animals to little or no profit, or turned into cider that did not sell for so much as it cost, counting the Apples of no value. Living immediately on a railroad that runs into this City, wherefrom my place is 35 miles distant, I should be able to do better with Apples than most growers; and yet I judge that half my Apples were of no use to me. Many of them sold in this City for $1 per barrel, including the cask, which cost me 40 cents; and, when you have added the cost of transportation, you can guess that

I had no surplus, after paying men $1.50 per day for picking and barreling them. I sold all I could to vinegar-makers at fifty cents per bushel for cider-apples—the casks being returned. But they could not take all I wished to sell them, there being so many sellers pressing to get rid of their windfalls before they rotted on their hands that even this market was glutted. That it was much worse for the farmer a dozen miles from a railroad and a hundred from the nearest city, none can doubt. I have heard that, in parts of Connecticut, cider was sold for fifty cents per barrel to whoever would furnish casks, and that their size was hardly considered. Manifestly, this left nothing for the apples.

If Apples could have been daily supplied to our poorer citizens in such quantities as they could conveniently take, at from fifty to seventy-five cents per bushel, according to quality and comeliness, I am confident that this City and its suburbs would have taken Two or Three Millions of bushels more than they have done; and the same is true of other cities. But the poor rarely buy a barrel of Apples at once; and they have been required to pay as much for half a peck as I could get for a bushel just like them. In other words: the hucksters and middlemen set so high a price on their respective services in dividing up a barrel of Apples and conveying them from the rural producer to the urban consumer that a large portion of the farmer's apples must rot on his hands or be sold by him for less than the cost of harvesting, while the poor of the cities find them too dear to be freely eaten.

Nor are Apples singular in this respect. I would like to grow a thousand bushels of English (round) and French or Swede Turnips per annum if I could be sure of getting $1 per barrel for them delivered at the railroad. If the poor of this City could buy such Turnips throughout their season by the half peck at the rate of $2 per barrel, I believe they would buy and eat many more than they do. But they are usually asked twenty-five cents per half peck, which is at the rate of $5 per barrel; and at this rate they hold them too dear for every-day use. So the Turnips are not grown, or the cattle are invited to clear them off before they rot and become worthless and a nuisance.

Quite often, a green youth undertakes to get rich by farming near some great city. He has heard and believes that Cabbages bring from $5 to $8 and even $10 per hundred, Squashes from $10 to $25 per hundred, Watermelons from $20 to $50, and so on. He has made his calculations on this basis, and sanguinely expects to make money rapidly. But his products, in the first place, fall short of his estimates; they are not ready for market so soon as he expected they would be; and, when at length they are ready, every one else seems to have rushed in ahead of him. The market is glutted; no one seems to want his "truck" at any figure; he sells it for a song, and quits farming disgusted and bankrupt. May be, his stuff would

have sold much better next week or the week after ·
but he could not afford to bring it to market and take
it back day after day, on the chance that the demand
for it would improve by-and-by.   I judge that more
young men have on this account turned their backs
on farming, after a brief trial, than on any other.
They might have borne up against the shortness of
their crops, hoping for better luck next time ; but
the necessity for selling them for a price that would
not have reïmbursed their cost, had they been ever so
luxuriant, utterly disheartens and alienates them.

I preach no crusade against hucksters and middlemen.   I hold them, in the actual state of things,
benefactors to both producers and consumers.   In so
far as they deal honestly and meet promptly their
obligations, they deserve commendation rather than
reproach.   What I urge is, that more economical and
efficient machinery of exchange and distribution
ought to be devised and set at work—machinery that
would do all that is required at a moderate, reasonable cost.

I would like to see one of our solvent, well-man-
aged Railroads advertise that it would henceforth buy
at any of its stations all the farmers' produce that
might be offered, and pay the highest prices that the
state of the markets would justify.   Let its agents
purchase whatever came along—a basket of eggs, a
coop of chickens, a barrel of apples, a sack of beans,
a pail of currants—anything that could be sold in
the city to which it runs, and which would conduce

to human sustenance or comfort. Its object should be Freight—the rapid and vast increase of its transportations, not extra profit on the articles transported. But let its agents be ready to buy at fair prices whatever was offered, paying cash down, and pushing everything purchased directly into market, so as to have the money back to buy more with directly. The Railroad Company, thus owning nearly everything edible it brought into market, would buy and sell at uniform prices, and not bid against itself, as a crowd of hucksters and middlemen will often do. I am confident that a Railroad that would inaugurate this system on a right basis, saying to every farmer living near it, " Grow whatever your soil is best adapted to, and bring it to our station : there, you shall have cash down for it, at the highest price we can afford to give," would rapidly double and quadruple its freights, and would thus build up a business which has no parallel under the present system.

It is urged, in opposition to this proposal, that a Railroad so managed would monopolize markets, and deal on its own terms with the producer and consumer. If there were but one railroad entering a great city, and no other mode of reaching it, this objection would be plausible, but not in the actual case. Whoever chose would be at liberty to start an opposition, and to use the railroad or dispense with it as he found advisable.

LI.

THE dearth of employment in Winter for farm laborers is a great and growing evil. Thousands, being dismissed from work on the farms in November, drift away to some city, under a vague, mistaken impression that there must be work at some rate where so much is being done and so many require service, and squander their means and damage their morals in fruitless quest of what is not there to be had. When Spring at length arrives, they sneak back to the rural districts, ragged, penniless, debauched, often diseased, and every way deteriorated, by their Winter plunge. For their sakes not only, but for the sakes also of those who will employ and those who must work with them hereafter, this drifting to the cities should be stopped.

In its present magnitude, it is a very modern evil. Far within my recollection, there was timber to cut and haul to the saw-mill, wood to cut, draw, and prepare for the year's fuel, with forest-land to be cleared and fitted for future cultivation, even in New-Eng-

(303)

land. Those who chose to work with ax or team were seldom idle in Winter. Now, there is little timber to cut, little land to clear, and coal is rapidly supplanting wood as fuel. So a larger and larger number of farm laborers is annually turned off when the ground freezes to live as they may for the next three or four months.

I recognize the right of the farmer, who has given twelve or more hours per day to the tillage of his acres and the saving of his crops throughout the genial months, to take the world more easily in Winter. He should now have leisure to return visits, to post and balance his books, and to improve his mind by study and reflection. Having worked hard when he must, he ought to rest and recuperate when he can. But he gravely errs who supposes that, the ground being frozen, there is no longer work to be done on the farm until the ground is fit to plow a-gain. On the contrary, he who realizes that the farmer is a manufacturer of food and fibrous substances from raw materials of far inferior value must see that, so soon as one harvest has been secured, the cultivator should devote his attention to the collec-tion and utilization of the elements wherefrom a larger crop may be obtained from the same acres next season.

And first as to Muck. No one who has not valued and sought it is likely to know how generally abun-dant and accessible this material is. I have found it in inexhaustible supply on the land of a pretty good cultivator who, after working a fair farm ten years,

sold it because (as he supposed) it was destitute of
this basis of extensive fertilization. "Seek, and ye
shall find," implies that those who do not seek will
rarely find; and such is the fact. Where rock
abounds, Muck is rarely wanting. It covers many
thousand acres of Jersey sands, where rock is unknown; but show me a region ridged or ribbed with
rock, and I shall confidently expect to find Muck on
it, though none has been known or supposed to exist
there. And he who either has or can buy a bed of
Muck within half a mile of his barn, his sty, his hen-
house, may dig and draw from it all Winter with a
moral certainty that it will generously reward his outlay. Begin as soon after haying as you can spare the
time, and cut an outlet so deep that you may thereafter work dryshod; thenceforth, dig and pile on the
nearest accessible spot of dry ground, to be drawn
away to the barn-yard and out-houses as opportunity
presents itself. But, even though you have done
nothing till the ground freezes, do not say it is now
too late, but set to work. You can often team in
Winter where you could not at any other season;
and, in digging Muck from a swamp or bog well frozen over, you are not apt to be troubled with water.
Draw all you can; but dig much more; for no money at lawful interest pays so well as Muck left to dry
and cure for months before you draw it. I think I
do not over-estimate the average value of a cubic yard
of Muck, well cured and mixed with warmer fertilizers before application to the soil, at one dollar; and

I think there are few farmers in the Old Thirteen States who cannot obtain it for less than that.

Where Muck is not to be had, I believe the tiller of a sandy or gravelly farm who can get access to a bed or bank of clay may profitably dig and draw this, to be used as he would use Muck if he had it, and even for direct application to the soil. I do not think this method the most advisable; yet I feel sure that clay spread over a sandy or gravelly field that has been laid down to grass is worth fifty cents per cubic yard wherever Hay is worth $12 per tun; but I would wish to apply it not later than December.

He who has fit places of deposit should draw all his Lime, Plaster, and other commercial fertilizers, in Winter, so as to be ready for use when required. Mix your Lime while fresh from the kiln with Muck, at the rate of a bushel of the former to a cubic yard of the latter, and the Muck will be ready for use far sooner than it otherwise would be. Be careful *not* to mix Lime with animal manures in any case, since it expels Ammonia, whereas the sulphur of Plaster combines with that volatile element and fixes it. There are some farmers who do, but twenty times as many who do not, use Plaster enough about their stables and pig-pens. They ought to realize that a bad smell implies a waste of Ammonia, which a farmer, unless very rich, can hardly afford.

Fences should all be scrutinized as Winter goes off, and put into thorough condition for next season's service.

Fruit-trees should be relieved of all dead or dying branches, all suckers, and cut back where towering too high, or spreading too wide. It may be better for the trees to do all pruning in May or June; but the farmer who defers it to that season is very likely to be hurried into postponing it to another year—and another.

There is scarcely a forest of second or later growth which would not pay for thinning and trimming, if well done. That which is cut out may be turned to good account as bean-poles, pea-brush, Summer fuel, etc., while that which is left will grow faster, taller, and more shapely, to reward you doubly for your pains.

—These are but suggestions. Any farmer can add to or improve upon them if he will give an hour's thought to the subject. The best laborers can be hired for a full year at a price not very much exceeding that which will secure their services for eight or nine months. In the interest alike of good crops and good morals, I urge every one who can to resolve that he will henceforth hire by the year, or in some way manage to employ his laborers in Winter as well as in Summer.

LII.

In the foregoing essays, I have set forth, as clearly as I could, the facts within my knowledge which seem calculated to cast light upon the farmer's vocation, and the principles or rules of action which they have suggested to my mind. I have been careful not to throw any false, delusive halo over this indispensable calling, and by no means to induce the belief that the farmer's lot is necessarily and uniformly a happy one. I know that his is not the royal road to rapid acquisition, and that few men are likely to amass great wealth by quietly tilling the soil. I know, moreover, that what passes for farming among us is not so noble, so intellectual, so attractive, a pursuit as it might and should be—that most farmers might farm better and live to better purpose than they do. Of all the false teaching, I most condemn that which flatters farmers as though they were demigods and their calling the grandest and the happiest ever followed by mortals, when the hearer, unless very green, must feel that the speaker does n't believe one word of all he utters; for, if he did, he

(308)

would be farming, instead of living by some profession, and talking as though his auditors did not know wheat from chaff. I regard the Agriculture of this country as very far below the standard which· it should ere this have reached : I hold that the great mass of our cultivators might and should farm better than they do, and that better farming would render their sons better citizens and better men. If a single line of this little work should seem calculated to cajole its readers into self-complacency rather than instruct them, I beg them to believe that their impression wrongs my purpose.

I am fully aware that others have treated my theme with fuller knowledge and far greater ability than I brought to its discussion. "Then why not leave them the field?" Simply because, when all have written who can elucidate my theme, at least three-fourths of those who ought to study and ponder it will not have read any treatise whatever upon Agriculture—will hardly have yet regarded it as a theme whereon books should be written and read. And, since there may be some who will read this treatise for its writer's sake—will read it when they could not be persuaded to do like honor to a more elaborate and erudite work—I have written in the hope of arousing in some breasts a spirit of inquiry with regard to Agriculture as an art based on Science —a spirit which, having been awakened, will not fall again into torpor, but which will lead on to the perusal and study of profounder and better books.

In the foregoing essays, I have sought to establish the following propositions:

1. That *good* farming is and must ever be a paying business, subject, like all others, to mischances and pull-backs, and to the general law that the struggle up from nothing to something is ever an arduous and almost always a slow process. In the few instances where wealth and distinction have been swiftly won, they have rarely proved abiding. There are pursuits wherein success is more envied and dazzling than in Agriculture; but there is none wherein efficiency and frugality are more certain to secure comfort and competence.

2. Though the poor man must often go slowly, where wealth may attain perfection at a bound, and though he may sometimes seem compelled to till fields not half so amply fertilized as they should be, it is nevertheless inflexibly true that bounteous crops are grown at a profit, while half and quarter crops are produced at a loss. A rich man may afford to grow poor crops, because he can afford to lose by his year's farming, while the poor man cannot. He ought, therefore, to till no more acres than he can bring into good condition—to sow no seed, plow no field, where he is not justified in expecting a good crop. Better five acres amply fertilized and thoroughly tilled than twenty acres which can at best make but a meager return, and which a dry or a wet season must doom to partial if not absolute failure.

3. In choosing a location, the farmer should resolve

to choose once for all. Roaming from State to State, from section to section, is a sad and far too common mistake. Not merely is it true that "The rolling stone gathers no moss," but the farmer who wanders from place to place never acquires that intimate knowledge of soil and climate which is essential to excellence in his vocation. He cannot read the clouds and learn when to expect rain, when he may look for days of sunshine, as he could if he had lived twenty years on the same place. Choose your home in the East, the South, the Center, the West, if you will (and each section has its peculiar advantages); but choose once for all, and, having chosen, regard that choice as final.

4. Our young men are apt to plunge into responsibilities too hastily. They buy farms while they lack at once experience and means, incur losses and debts by consequent miscalculations, and drag through life a weary load, which sours them against their pursuit, when the fault is entirely their own. No youth should undertake to manage a farm until after several years of training for that task under the eye of a capable master of the art of tilling the soil. If he has enjoyed the requisite advantages on his father's homestead, he may possibly be qualified to manage a farm at twenty-one; but there are few who might not profitably wait and learn, in the pay of some successful cultivator, for several years longer; while I cannot recall an instance of a youth rushing out of school or a city counting-house to show old farmers

how their work ought to be done, that did not result
in disaster. It is very well to know what Science
teaches with regard to farming; but no man was
ever a thoroughly good farmer who had not spent
some years in actual contact with the soil.

5. While every one says of his neighbor, "He
farms too much land," the greed of acquisition does
not seem at all chastened. Men stagger under loads
of debt to-day, who might relieve themselves by
selling off so much of their land as they cannot
profitably use; but every one seems intent on holding all he can, as if in expectation of a great advance
in its market value. And yet you can buy farms in
every old State in the Union as cheaply per acre as
they could have been bought in like condition sixty
years ago; and I doubt their selling higher sixty
years hence than they do now. No doubt, there *are*
lands, in the vicinage of growing cities or villages,
that have greatly advanced in value; but these are
exceptions: and I counsel every young farmer, every
poor farmer, to buy no more land than he can cultivate thoroughly, save such as he needs for timber.
Never fear that there will not be more land for sale
when you shall have the money wherewith to buy it;
but shun debt as you would the plague, and prefer
forty acres all your own to a square mile heavily
mortgaged. I never lifted a mill-stone; but I have
undertaken to carry debts, and they are fearfully
heavy.

6. I know that most American farms east of the

Roanoke and the Wabash have too many fields and
fences, and that the too prevalent custom of allowing
cattle to prowl over meadow, tillage and forest, from
September to May, picking up a precarious and inadequate subsistence by browsing and foraging at
large, is slovenly, unthrifty, and hardly consistent
with the requirements of good neighborhood. It is
at best a miseducation of your cattle into lawless
habits. I do not know just where and when *all* pasturing becomes wasteful and improvident; but I do
know that pasturing fosters thistles, briers, and every
noxious weed, and so is inconsistent with cleanly and
thorough tillage. I know that the same acres will
feed far more stock, and keep them in better condition, if their food be cut and fed to them, than if they
are sent out to gather it for themselves. I know that
the cost of cutting their grass and other fodder with
modern machinery need not greatly exceed that of
driving them to remote pastures in the morning and
hunting them up at nightfall. I know that penning
them ten hours of each twenty-four in a filthy yard,
where they have neither food nor drink, is unwise;
and I feel confident that it is already high time,
wherever good grass-land is worth $100 per acre, to
limit pasturage to one small field, as near the center
of the farm as may be, wherein shade and good
water abound, into which green rye, clover, timothy,
oats, sowed corn, stalks, etc., etc., may successively
be thrown from every side, and where shelter from a
cold, driving storm, is provided; and that, if cows

14

could be milked here and left through night as well as day, it would be found good economy.

7. I know that most of us are slashing down our trees most improvidently, and thus compelling our children to buy timber at thrice the cost at which we might and should have grown it. I know that it is wasteful to let White Birch, Hemlock, Scrub Oak, Pitch Pine, Dogwood, etc., start up and grow on lands which might be cheaply sown with the seeds of Locust, White Oak, Hickory, Sugar Maple, Chestnut, Black Walnut, and White Pine. I know that no farm in a settled region is so large that its owner can really afford to surrender a considerable portion of it to growing indifferent cord-wood when it would as freely grow choice timber if seeded therefor; and I feel sure that there are few farms so small that a portion of each might not be profitably devoted to the growing of valuable trees. I know that the common presumption that land so devoted will yield no return for a life-time is wrong—know that, if thickly and properly seeded, it will begin to yield bean-poles, hoop-poles, etc., the fifth or sixth year from planting, and thenceforth will yield more and more abundantly forever. I know that *good* timber, in any well-peopled region, should not be *cut off*, but *cut out*— thinned judiciously but moderately and trimmed up, so that it shall grow tall and run to trunk instead of branches; and I know that there are all about us millions of acres of rocky crests and acclivities, steep ravines and sterile sands, that ought to be seeded to

timber forthwith, kept clear of cattle, and devoted to tree-growing evermore.

8. I do not know that all lands may be profitably underdrained. Wooded uplands, I know, could not be. Fields which slope considerably, and so regularly that water never stagnates upon or near their surface, do very well without. Light, leachy sands, like those of Long Island, Southern Jersey, Eastern Maryland, and the Carolinas, seem to do fairly without. Yet my conviction is strong that *nearly all land which is to be persistently cultivated will in time be underdrained.* I would urge no farmer to plunge up to his neck into debt in order to underdrain his farm. But I *would* press every one who has no experience on this head to select his wettest field, or the wettest part of such field, and, having carefully read and digested Waring's, French's, or some other approved work on the subject, procure tile and proceed next Fall to drain that field or part of a field thoroughly, taking especial precautions against back-water, and watch the effect until satisfied that it will or will not pay to drain further. I think few have drained one acre thoroughly, and at no unnecessary cost, without being impelled by the result to drain more and faster until they had tiled at least half their respective farms.

9. As to Irrigation, I doubt that there is a farm in the United States where *something* might not be profitably done forthwith to secure. advantage from the artificial retention and application of water.

Wherever a brook or runnel crosses or skirts a farm, the question—" Can the water here running uselessly by be retained, and in due season equably diffused over some portion of this land ?"—at once presents itself. One who has never looked with this view will be astonished at the facility with which some acres of nearly every farm may be irrigated. Often, a dam that need not cost $20 will suffice to hold back ten thousand barrels of water, so that it may be led off along the upper edge of a slope or glade, falling off just enough to maintain a gentle, steady current, and so providing for the application of two or three inches of water to several acres of tillage or grass just when the exigencies of crop and season most urgently require such irrigation. Any farmer east of the Hudson can tell where such an application would have doubled the crop of 1870, and precluded the hard necessity of selling or killing cattle not easily replaced.

Of course, this is but a rude beginning. In time, we shall dam very considerable streams mainly to this end, and irrigate hundreds and thousands of acres from a single pond or reservoir. Wells will be sunk on plains and gentle swells now comparatively arid and sterile, and wind or steam employed to raise water into reservoirs whence wide areas of surrounding or subjacent land will be refreshed at the critical moment, and thus rendered bounteously productive. On the vast, bleak, treeless Plains of the wild West, even Artesian wells will be sunk for this purpose; and

the water thus obtained will prove a source of fertility as well as refreshment, enriching the soil by
the minerals which it holds in solution, and insuring
bounteous crops from wide stretches of now barren
and worthless desert. Immigration will yet thickly
dot the great Sahara with oases of verdure and plenty;
but it will, long ere that, have covered the valleys
of our Great Basin and those which skirt the affluents of the savage and desolate Colorado with a
beauty and thrift surpassing the dreams of poets.
And yet, its easiest and readiest triumphs are to be
won right here—in the valleys of the Connecticut,
the Hudson, the Susquehanna, and the Potomac.

10. As to Commercial Fertilizers, I think I have
been well paid for the application of Gypsum (Plaster
of Paris) to my upland grass at the rate of one bushel
per acre per annum, while my tillage has been supplied with it by dusting my stables with it after each
cleaning, and so applying it mingled with barn-yard
manures. Lime (unslaked) from burned oyster-shells,
costing me from 25 to 30 cents per bushel delivered,
I have applied liberally, and I judge, with profit.
Bones, ground, (the finer the better) I have largely
and I think advantageously used; but my land had
been mainly pastured for nearly two centuries before
I bought it, and thus continually drained of Phosphates, yet never replenished: so my experience does
not prove that the farmers of newer lands ought to
buy bones, though I advise them to apply all they
can save or pick up at small cost. Pound them very

fine with a beetle or ax-head on a flat stone, and give them to your fowls: if they refuse a part of them, your soil will prove less dainty. I am not sure that it pays to buy any manufactured Phosphate when you can get Raw Bone; though I doubt not that, for instant effect, the Phosphate is far superior. As to Guano, it has not paid me; but that may be the fault of careless or unskillful application. I judge that any one who has to deal with sterile sands that will not bring Clover, may wisely apply 400 pounds of Guano per acre, provided he has nothing else that will answer the purpose. After he has produced one good stand of Clover, I doubt that he can afford to buy more Guano, unless he can apply it to better purpose than I have yet done.

I have a strong impression that most farmers can do better at making and saving fertilizers than by buying them. Lime and Sulphur (Gypsum), if your soil lacks them, you must buy; but a good farmer who keeps even a span of horses, three or four cows, as many pigs, and a score of fowls, can make for $100 fertilizers which I would rather have than two tuns of Guano, costing him $180 to $200. If he has a patch of bog or a miry pond on his farm—any place where frogs will live—he can dig thence, in the dryest time next Fall, two or three hundred loads of Muck, which, having been left to dry on the nearest high ground till November or later, and then drawn up and dumped into his barn-yard, pig-pen, and fowl-house, will be ready to come out next Spring in season for corn-planting, and, being liberally applied, will do as much for his crop as two tons of Guano would, and will strengthen his land far more. If he has no Muck, and no neighbor who can spare it as well as not, let him at midsummer cut all the weeds growing on and around his farm, and in the Fall gather all the leaves that can be impounded, using these as litter for his cattle and beds for his pigs, and he will be agreeably surprised at the bulk of his heap next Spring.

I am an intense believer in Home Production. We send ten thousand miles for Guano, and suffer the equally valuable excretions of our cities to run to waste in rivers and bays, poisoning or driving away the fish, and filling the air with stench and pestilence. No farmer ever yet intelligently *tried* to enrich his land and was defeated by lack of material. He may not be able to do all he would like to at first; but persistent effort cannot be baffled.

11. Shallow culture is the most crying defect of our average farming. Poverty may sometimes excuse it; but the excuse is stretched quite too far. If a farmer has but a poor span of horses, or a light yoke of thin steers, he cannot plow land as it should be plowed; but let him double teams with his neighbor, and plow alternate days on either farm; or, if this may not be, let him buy or borrow a sub-soil plow, and go once around with his surface plow, then hitch on to the sub-scil, and run another furrow in the bot-

tom of the former. There are a few intervales of
rich, mellow soil, deposited by the inundations of
countless ages, where shallow culture will answer,
because the roots of the plants run freely through
fertile earth never yet disturbed by the plow; but
these marked and meagre exceptions do not invali-
date the truth that nine-tenths of our tillage is
neither so deep nor so thorough as it should be. As
a rule, the feeding-roots of plants do not run below
the bottom of the furrows, though in some instances
they do; and 'he who fancies that five or six inches
of soil will, under our fervid suns, with our Summers
often rainless for weeks, produce as bounteous and as
sure a crop as twelve to eighteen inches, is impervious
to fact or reason. He might as sensibly maintain
that you could draw as long and as heavily against
a deposit in bank of $500 as against one of $1,500.

12. Finally, and as the sum of my convictions, we
need more thought, more study, more intellect, in-
fused into our Agriculture, with less blind devotion
to a routine which, if ever judicious, has long since
ceased to be so. The tillage which a pioneer, fighting single-handed and all but empty-handed with a
dense forest of giant trees, which he can do no better
than to cut down and burn, found indispensable
among their stumps and roots, is not adapted to the
altered circumstances of his grandchildren. If our
most energetic farmers would abstract ten hours each
per week from their incessant drudgery, and devote
them to reading and reflection with regard to their

noble calling, they would live longer, live to better purpose, and bequeath a better example, with more property, to their children.

———

My self-imposed task is done. I undertook to tell What I Know of Farming through one brief essay for each week in 1870; and, in the face of multifarious and pressing duties, and in despite of a severe, protracted illness, the work has been prosecuted to completion. Had I not kept ahead of it while in health, there were weeks when I must have left it unaccomplished, as I was too ill to write or even stand.

I close with the avowal of my joyful trust that these essays, slight and imperfect as they are, will incite thousands of young farmers to feel a loftier pride in their calling and take a livelier interest in its improvement, and that many will be induced by them to read abler and better works on Agriculture and the sciences which minister to its efficiency and impel its progress toward a perfection which few as yet have even faintly foreseen.

INDEX.

THE END.

Horace Greeley's Autobiography.

RECOLLECTIONS OF A BUSY LIFE

INCLUDING

REMINISCENCES OF AMERICAN POLITICS AND POLITICIANS,

From the Opening of the Missouri Contest to the Downfall of Slavery.

By HORACE GREELEY.

In one elegant octavo volume. Beautifully printed and handsomely bound. Illustrated with a fine *Steel Portrait of Mr. Greeley*, also with Wood Engravings of "The Cot where I was Born," "My First School House," "Portrait of Margaret Fuller," "My Evergreen Hedge," "My House in the Woods," "My Present Home," "My Barn."

DEDICATED
TO
OUR AMERICAN BOYS,
WHO,

BORN IN POVERTY, CRADLED IN OBSCURITY, AND EARLY CALLED FROM SCHOOL TO RUGGED LABOR, ARE SEEKING TO CONVERT OBSTACLE INTO OPPORTUNITY, AND WREST ACHIEVE- MENT FROM DIFFICULTY,

THESE RECOLLECTIONS
ARE REGARDFULLY INSCRIBED BY
THEIR AUTHOR.

Mr. Greeley himself gives the best indication of their nature, when he says: "I shall never write anything else into which I shall put so much of *myself*, my experiences, notions, convictions, and modes of thought as these *Recollections*. I give, with small reserve, my mental history."

In his "Apology," Mr. Greeley says: "* * * If my friends will accept the essays which conclude this volume as a part of my mental biography, I respectfully proffer this book as my account of all of myself that is worth their consideration; and I will cherish the hope that some portion, at least, of its contents embody lessons of persistence and patience, which will not have been set forth in vain."

PRICES:—EXTRA CLOTH, $2 50. LIBRARY STYLE (sheep), $3 50. HALF MOROCCO, $4 00. HALF CALF, ELEGANT, $5 00. MOROCCO ANTIQUE, $7 00.

New-York Tribune.

1871.   DAILY, SEMI-WEEKLY AND WEEKLY.   1871.

THE WEEKLY TRIBUNE.
The Paper of the People.

THE TRIBUNE aims to be pre-eminently a *News*-paper. Its correspondents traverse every State, are present on every important battle-field, are early advised of every notable Cabinet decision, observe the proceedings of Congress, of Legislatures, and of Conventions, and report to us by telegraph all that seems of general interest. We have paid for one day's momentous advices from Europe by Cable far more than our entire receipts for the issue in which those advices reached our readers. If lavish outlay, unsleeping vigilance, and unbounded faith in the liberality and discernment of the reading public, will enable us to make a journal which has no superior in the accuracy, variety, and freshness of its contents, THE TRIBUNE shall be such a journal.

To Agriculture and the subservient arts, we have devoted, and shall persistently devote, more means and space than any of our rivals. We aim to make THE WEEKLY TRIBUNE such a paper as no farmer can afford to do without, however widely his politics may differ from ours. Our reports of the Cattle, Horse, Produce, and General Markets, are so full and accurate, our essays in elucidation of the farmers's calling and our regular reports of the Farmers' Club and kindred gatherings, are so interesting, that the poorest farmer will find therein a mine of suggestion and counsel, of which he cannot remain ignorant without positive and serious loss.

AS A FAMILY NEWSPAPER,

THE WEEKLY TRIBUNE is pre-eminent. In addition to Reviews, Notices of New Books, Poetry, &c., we publish Short Stories, original or selected, which will generally be concluded in a single issue, or at most in two or three. We intend that THE TRIBUNE shall keep in the advance in all that concerns the Agricultural, Manufacturing, Mining, and other interests of the country; and that, for variety and completeness, it shall remain altogether the most valuable, interesting and instructive NEWSPAPER published in the world.

No newspaper so large and complete as THE WEEKLY TRIBUNE was ever before offered at so low a price.

TERMS OF THE WEEKLY TRIBUNE.
To Mail Subscribers.

One copy, one year, 52 issues, $2.
Five copies....... ... .   9.

To ONE ADDRESS, all at one Post Office.
Ten Copies..........$1 50 each.,
Twenty Copies...... 1 25   "
Fifty Copies........ 1 00   "
And One Extra Copy to each Club.

To NAMES OF SUBSCRIBERS, all at one Post Office.
Ten Copies.........$1 60 each.
Twenty Copies...... 1 35   "
Fifty Copies........ 1 10   "
And One Extra Copy to each Club.

Persons entitled to an extra copy can, if preferred, have either of the following books, postage prepaid: Political Economy, by Horace Greeley; Pear Culture for Profit, by P. T. Quinn; The Elements of Agriculture, by Geo. E. Waring.

THE NEW YORK SEMI-WEEKLY TRIBUNE

is published every TUESDAY and FRIDAY. THE SEMI-WEEKLY TRIBUNE gives, in the course of a year, Three or Four of the

Best and Latest Popular Novels,

by living authors. Nowhere else can so much current intelligence and permanent literary matter be had at so cheap a rate as in THE SEMI-WEEKLY TRIBUNE.

TERMS OF THE SEMI-WEEKLY TRIBUNE.

One copy, one year—104 No's...$4 00
Two copies.....................   7 00
Five copies, or over, each copy.. 3 00

An extra copy will be sent for every club of ten sent for at one time; or, if preferred, a copy of Recollections of a Busy Life, by Mr. Greeley.

DAILY TRIBUNE.

Mail Subscribers......$10 per annum.

Address   **THE TRIBUNE, New York.**

Lightning Source UK Ltd.
Milton Keynes UK
UKOW05f0507190516

274557UK00015B/368/P